Palgrave Studies in Anthropology of Sustainability

Series Editors
Marc Brightman
Department of Anthropology
University College London
London, UK

Jerome Lewis
Department of Anthropology
University College London
London, UK

Our series aims to bring together research on the social, behavioral, and cultural dimensions of sustainability: on local and global understandings of the concept and on lived practices around the world. It publishes studies which use ethnography to help us understand emerging ways of living, acting, and thinking sustainably. The books in this series also investigate and shed light on the political dynamics of resource governance and various scientific cultures of sustainability.

More information about this series at
http://www.palgrave.com/gp/series/14648

Annelie Sjölander-Lindqvist
Ivan Murin • Michael E. Dove
Editors

Anthropological Perspectives on Environmental Communication

palgrave
macmillan

Editors
Annelie Sjölander-Lindqvist
School of Global Studies &
Gothenburg Research Institute
University of Gothenburg
Göteborg, Sweden

Ivan Murin
Department of Social Studies and
Ethnology
Matej Bel University
Banská Bystrica, Slovakia

Michael E. Dove
Department of English and
American Studies
Matej Bel University
Banská Bystrica, Slovakia

Palgrave Studies in Anthropology of Sustainability
ISBN 978-3-030-78039-5 ISBN 978-3-030-78040-1 (eBook)
https://doi.org/10.1007/978-3-030-78040-1

This Palgrave Macmillan imprint is published by the registered company Springer Nature Switzerland AG.
The registered company address is: Gewerbestrasse 11, 6330 Cham, Switzerland

Fig. 1 The authors' various field locations from a global perspective. (This image used with permission of Pavol Midula. Source of topographical data: ESRI. Program: ArcGIS 10.5, MAC Address: 2c27d72be320, Host Name: fpv-midula)

To all the people and places who make our research possible.

Foreword

Anthropology is not my academic home port, but I find it enormously compelling and visit often. I admire anthropologists because they are disciplined, observant, and inquisitive field researchers driven by a seemingly insatiable curiosity to unpack and better understand the distinctions that define a particular human community. With accuracy and respect, they seek to identify culture markers including the ancient, historic, and/or contemporary patterns, rituals, beliefs, relationships, and roles that define how particular humans have worked out how they will live together in their environment.

Perhaps it is the voyeur in me, in all of us, that draws me to anthropological writing. I am amazed with the endless diversity, resourcefulness, and resilience of how humans do life together with other humans in a particular place. Anthropologists nearly always introduce me to cultures I did not know existed and that I will likely never be able to physically visit.

The anthropologists in this volume do not disappoint. Geographically, they take us to communities in Sweden's western coastal area and a former mining area in the country's central region, rural areas of Chile and Central Slovakia, protected areas in Arizona and Utah in the southwestern USA, and the largest lake in Alaska.

Yet these anthropologists take us beyond their discipline's traditional foci as Sjölander-Lindqvist describes anthropology in her introduction to this book as "interested in human diversity and real-world contexts; what happens between people in certain settings, what drives the way people understand the surrounding world, and how people act within and wherever they happen to live."

That is where I come in. My academic home port is the Communication discipline, in particular Environmental Communication (EC). "At its foundation, Communication focuses on how people use messages to generate meanings within and across various contexts, and is the discipline that studies all forms, modes, media, and consequences of communication through humanistic, social scientific, and aesthetic inquiry" (National Communication Association, 2021).

As such, Communication researchers regularly venture out of our home port to interact with anthropology as well as psychology, political science, sociology, media, economics, public policy, rhetoric and persuasion, history, language, and cultural studies. Although this litany of interests might make Communication seem like it is all over the map, it is distinguished by its foundational beam on the dynamics of the Social Construction of Reality, or what is commonly called Symbolic Interactionism (Blumer, 1937). Communication cuts across contexts and situations; it is the relational and collaborative force that strategically constructs the social world.

We act toward others and non-humans based on the meanings that we assign to them, and this meaning is based on the social interaction that we have or do not have with others. Meanings are not inherent in the verbal, nonverbal, and visual messages; meanings are instead socially/mutually created and are not necessarily static (Lyman et al., 2001). Meanings might endure, or they might shift or be contested. The meanings that prevail or dominate become fundamental drivers of our actions, priorities, and decisions. Communication is interested in the *dynamics of meaning making* in every kind of human setting. We advocate truthfulness, accuracy, honesty, and reason as essential to the integrity of communication.

Environmental Communication (EC) is a distinct field within Communication. It is certainly multifaceted, as described on the websites of the NCA (Environmental Communication Division, 2021) and the International Environmental Communication Association (2021), comprising academics, professionals, and practitioners who are particularly interested in all communication aspects of Environmental/Natural Resource Management (ENRM). Like Communication, EC is interested in all settings and modes of messaging about the environment, but with an emphasis on improving human capacity to address ENRM challenges in productive ways toward justice and sustainability.

What follows is my version of EC, so please indulge my rambling, because I think it will help readers appreciate the chapters of this volume. Because humans mutually construct meaning by interacting with each other, attitudes and actions toward the natural world are determined by human relationships and messaging *with each other* about it. It is this social construction or mutual meaning making of humans toward the earth that captivates us. Consider this: Does it make any difference if your childhood bedtime book was *The Three Little Pigs and the Big Bad Wolf* (Jacobs & Batten, 1967) or *The Three Little Wolves and the Big Bad Pig* (Trivizas & Oxenbury, 1993)?

Particularly Western humans have mutually created over time a dualistic paradigm to describe the separate relationship between humans and the natural world. Because of seventeenth-century philosopher René Descartes' revolutionary and prevailing approach to logic, called Cartesian thought, traditional approaches to ENRM separate the brain from the body, emotions from reason, and for our purposes here, culture from biophysical sciences/technology. It elevates and prioritizes physical evidence, reason, and science as more important, valuable, worthy, and powerful than culture, emotion, heritage, or spirituality.

Over the centuries, Western decision making about ENRM situations has taken Cartesian thought quite literally, treating nature like a machine with expendable parts that can be adjusted or eliminated, and applying dichotomous properties of Cartesian thought to pretty much everything, mindlessly seeing and valuing the world through the dualistic lens of nature as separate from humans and their cultures and built environments.

This dualism influenced how government agencies were constructed as self-isolating and increasingly nonfunctional silos (Samson & Knopf, 2001). It has established biophysical/technological protocols for making ENRM decisions and, still most dominant today, humans with presumed impunity engage the natural world as the other, a machine. We act with unawareness of the damage this has caused, and continues to cause, to ecological systems including humans.

I vividly remember the first days of my three years as a U.S. Peace Corps Volunteer in the South Pacific country of Western Samoa (yes, I met elders who told me stories about Margaret Mead!). Village fishers had been taught by World War II (WWII) soldiers stationed there in the 1940s, who had lots of free time and lots of ammunition but no understanding of ecological systems, how to dynamite the reefs for easy fishing. The ocean and reefs held strong attachments in Samoan culture, but the fishing was easy and Samoan cultural experience did not include what dynamite would do to the reefs—until the damage was undeniable. When I was there in the late 1970s, many of the damaged or destroyed reefs were barely starting to recover, if at all, even after decades of generally enforced prohibitions against dynamiting. Village fishers had to travel beyond the reefs into the dangerous open ocean to catch fish, but most did not have outrigger boats capable of doing that. Instead they had to spend what little cash they had or barter for tinned fish from Australia. This resulted in fewer meals of fish, lower-protein diets, and the highest level of childhood malnutrition at that time among South Pacific countries. Sea turtles became scarcer as diets depended more on them for protein. At the request of the Samoan government, the Peace Corps was involved in sea turtle conservation projects as well as assisting the establishment of fishing cooperatives for traveling to the open ocean to obtain more sustainable protein while the reefs hopefully recovered.

It was that mutual construction of meaning among the WWII soldiers and Samoans that determined how they chose to interact with the reefs, and it was human–human messaging that later challenged this choice and shifted behavior based on a more holistic understanding of themselves and the reef as one mutually interdependent system.

It is deconstructing, critiquing, and understanding the dynamics of this meaning making that is the core of EC, because the outcomes of collaborative, persuasive, and contentious human communication processes determine local to global policies, processes, behaviors, and actions that have significant consequences for the planet's sustainability and human communities.

While the Samoans were organizing fishing cooperatives and hoping their dynamited reefs would recover, Kenyan Wangari Maathai was starting a small tree nursery in her backyard to address the devastating social and ecological effects of deforestation and desertification. Her efforts launched the grassroots organization Kenya's Green Belt Movement, composed mostly of women working to grow and plant tree seedlings to improve soil, hold rainwater, provide firewood, and grow food. In 2004, with 5000 grassroots nurseries operating and over 20 million trees planted throughout Kenya, Wangari Maathai was awarded the Nobel Peace Prize. *This authoritative body's recognition was extraordinary; global security was explicitly linked to the consequences of human behavior for ecological health and sustainability.* In 2007, the Nobel Committee reinforced this by awarding the Nobel Peace Prize to the United Nations Intergovernmental Panel on Climate Change and former U.S. Vice President Al Gore. Gore's 2006 documentary *An Inconvenient Truth* has received too many international awards to list and is a common part of school curriculums around the world. Think of the verbal and visual messaging conundrums the filmmakers had to overcome in order to reach diverse audiences, each with their own sense of who they are in a particular place!

And this is the intersection where the authors of this book present their cases, where Anthropology and Environmental Communication meet. Each case takes us to a distinct community and identifies that community's particular ways of being through the anthropological lens and additionally gazes through an EC lens to present the roles that each community's sense of self-in-place (Cantrill & Senecah, 2001) plays in the ENRM dilemma confronting them. Most importantly for EC, the cases reveal the dynamics of the struggles to determine which socially constructed meanings and their symbolic representations will prevail and integrally define, preserve, or reinforce who these communities and stakeholders believe themselves to be in this place at this time. In turn, those embraced identities drive their perspectives and behavior choices when faced with an ENRM dilemma. Stories are prominently featured as powerful identity creators and holders, as are personal and collective memories, cultural practices, place names, land practices, and relationships with landscapes.

Each community's dilemma is fueled by a tension or competition between the dualism of culture/heritage and biophysical science/technology. These kinds of tensions/competitions are also a key component of EC. Dichotomous epistemologies often compete to influence ENRM processes and decisions.

This tension or competition erupts from the complexity, and often contentiousness, that defines ENRM situations. Their complexity often comes from the need to satisfy multiple objectives, such as biophysical, economic, recreational, health, or social needs. They are nearly always beholden to legal and/or policy frameworks, subject to financial constraints, faced with uncertain scientific and technological knowledge, bound to timelines.

The contentiousness often comes from the choice or requirement to involve multiple stakeholders who typically have different priorities, different perspectives, different attachments, uneven authority, uneven resources, or uneven process experience. These processes and campaigns often get messy and can be mentally and emotionally exhausting, but if they are done well, they are necessary, worthwhile, and satisfying, because how a community makes ENRM decisions is a key component of their cultural and ecological sustainability.

And this brings me back to the chapters of this book that introduce us to communities who fear losing their identity, fear losing who they are in these places, their integral attachments, their sense of self-in-place. Their fears are about losing culture, landscapes, language, oral histories, symbols of heritage, agricultural practices, access to water, and so on. Losing themselves. Keep this in mind.

Alarcón in the chapter "Power, Conflicts, and Environmental Communication in the Struggles for Water Justice in Rural Chile: Insights from the Epistemologies of the South and the Anthropology of Power" especially demonstrates the complexity and contentiousness of ENRM decision-making processes, in this case to address Chile's water crisis and the conflict between poorer communities and private water holders. The struggle for water justice in Chile focuses on the poorer communities' perceived lack of voice (Senecah, 2004) in the government's water roundtable, revealing a deep distrust and power disparity in communities' struggle for equitable water access.

In the chapter "'The Sea Has No Boundaries': Collaboration and Communication Between Actors in Coastal Planning on the Swedish West Coast", Larsson and Sjölander-Lindqvist deconstruct a complex and contentious but largely productive decision-making process for coastal zone planning on Sweden's West Coast. They focus on boundaries and the dilemmas they create for the process and its stakeholders, for example, boundaries of roles, responsibilities, knowledge, laws, and jurisdictions. Whether you read this chapter as an academic researcher, a practitioner, a "pracademic" like me, or a stakeholder, this case captures how exhausting collaborative processes can be, even when they are very much worth the effort.

In the chapter "Arsenic Fields: Community Understandings of Risk, Place, and Landscape" by Sjölander-Lindqvist, the communities of the copper mining fields in Riddarhyttan of central Sweden cherish their strong, historic, and cultural attachments to their land, including the positive role that the copper mines played in creating them. Their dilemma is how to reconcile the risks of staying with a dubious proposal for arsenic remediation if the other option is to leave. The risk assessment process highlights the necessity of relationship trust in terms of information access, legitimacy of diverse perspectives, and influence on the decision—what I call the Trinity of Voice (Senecah, 2004).

In the chapter "Community Voices, Practices, and Memories in Environmental Communication: Iliamna Lake Yup'ik Place Names,

Alaska", Kugo gives us an insider's view and appreciation for the powerful meanings of the Yup'ik people's place names for the Iliamna Lake area in Southwest Alaska. The indigenous Yup'ik's oral traditions, communal stories, historic narratives, and the richness of these place names define their sense of self-in-place. Their dilemma is to resist the colonization of place names that causes meaning to fade, and preserve not only these place names, but even more importantly, the treasured associations and attachments these names hold, so that Yup'ik identity will endure. Kugo's chapter is also impressive in the methodology described to earn the trust necessary to enter this community.

Van Vlack in the chapter "Dancing with Lava: Indigenous Interactions with an Active Volcano in Arizona" describes a similar dilemma of indigenous meaning and attachments to a landscape now located in a national park. The Southern Paiute tribes' oral history and archeologists' publications are at heritage-interpretation odds over how the Little Springs Lava Flow in northern Arizona was used in the far past. Van Vlack details the ancient Paiute understandings of this landscape and suggests that the U.S. National Park Service's trend toward showcasing multiple voices of interpretation might be a positive option for preserving and respecting these dueling epistemologies.

Stoffle's case (chapter "Living Stone Bridges: Epistemological Divides in Heritage Environmental Communication") presents the dilemmas that arise when a U.S. National Park does what Van Vlack suggests and actively features multiple interpretative voices. The relationship between several tribes and pueblos and the Living Stone Bridge in northern Arizona is one of family relations. To the tribes and pueblos, the stone formation is a living entity, a relative, not a cultural symbol. The indigenous peoples did not socially construct its meaning; it is, in and of itself, alive, as is everything. Because to the indigenous peoples it is a separate living thing and because the Park Service wants to honor that, Stoffle poses the dilemma of how might the Living Stone Bridge be included in the decision-making process if it were truly recognized as a stakeholder?

What an intriguing question! The U.S. legal system has allowed humans to have standing on behalf of nature for some time, stemming from the Hudson River Storm King case in 1969. But how to establish or operationalize actual legal personhood for non-humans, and how would they represent themselves? We might look to some examples of legal personhood and the humans assigned as their guardians and representatives. The Shinnecock Nation of New York State has established legal personhood for water (Leonard, 2020). New Zealand's Parliament conferred legal personality on the Whanganui River with the same rights and responsibilities as a person to reflect the Whanganui iwi's unique ancestral relationship with the river (New Zealand, 2017). The Yurok Tribe of California declared rights of personhood for the Klamath River in tribal courts. Stoffle, Van Vlack, and Kugo's cases provide much to think about in terms of how to honor and reconcile diverse worldviews in interpretation and in legal systems.

The chapters "Cultural Transmission in Slovak Mountain Regions: Local Knowledge as Symbolic Argumentation" and "Demographic Change and Local Community Sustainability: Heritagization of Land Abandonment Symbols" take us to Central Slovakia, where, because of demographic, economic, and land use changes, communities fear the loss of their identity and cultural heritage. Murin takes us to the Hillside Settlements, where the dwindling population of long-time residents is intent on preserving what are now abandoned agricultural practices and attachments to cultural landscapes. Murin, Horský, and Aláč introduce us to a similar community concerned with preserving abandoned symbols of their cultural heritage, particularly cemetery markers. The authors apply the concept of "heritagization" to describe the process of reclaiming these kinds of abandoned land symbols to rediscover and reenergize cultural memory and identity.

As noted earlier, this paradigm of dualism in ENRM decision making is slowly and deliberately changing to more holistic understanding of humans and their built environment being critical parts of any ecosystem (McLeod & Leslie, 2009). Humans and human culture are still generally not recognized as having a place in any ecological system. In fact, until recent decades, the natural world was not recognized at all as a system or web of mutually interdependent parts.

In 1892, Ellen Swallow (who was the first woman to complete doctorate studies in chemistry, but as a woman was not allowed the Ph.D. degree or the title of Doctor) was the first to use the term "ecology," based on German zoologist Ernst Haeckel's concept of "Oekology," as a science focusing on the interrelationship between organisms and their environments, the "economy of nature," or literally "nature's household" (Lawrence, 2001, p. 675). Dr. Swallow's "ecology" neatly captured her broad concerns for human-created environmental problems, and her work was influential in improved sanitary conditions in urban areas. Central to her concept of ecology was that the environment formed people's lived experiences and that they responded to it both physically, in terms of their health and well-being, and socially, in that they came to accept as normal what was in fact constructed by societal arrangements and policy process. Conditions could be improved if people only knew about and agitated for that change, insisting, "the environment that people live in is the environment that they learn to live in, respond to, and perpetuate" (Clarke, 1973, p. 159). I think if EC had existed then, Dr. Swallow would have embraced it!

As Dr. Swallow posited, socially constructed normalities can change as a result of human experience and agitation. Aldo Leopold, often called the "father of ecology" in terms of wild/non-urban ecology and another EC hero of mine, also recognized how wrong-headed Descartes' dualistic thinking was and influenced the early shifts of heavy, dominating Cartesian thought. As a U.S. Forest Service employee in New Mexico in 1912, Leopold shot a wolf from the top of a rimrock canyon in New Mexico (we can guess which wolf book he read as a child). He reached the still-breathing wolf and saw something that forever changed him. In his classic text, *A Sand County Almanac*, Leopold describes the experience:

> In those days we had never heard of passing up the chance to kill a wolf. In a second we were pumping lead into the pack but with more excitement than accuracy. [...] We reached the old wolf in time to watch a fierce green fire dying in her eyes. I realized then, and have known ever since, that there was something new to me in those eyes—something known only to her and to the mountain. I was young then, and full of trigger-itch; I thought that because fewer wolves meant more deer, that no wolves would mean hunters' paradise. But after seeing the green fire die, I sensed that neither the wolf nor the mountain agreed with such a view. (Leopold, 1969, p. 130)

When wolves were nearly extirpated and deer subsequently overpopulated, and the available habitat grew smaller and sickly, then died of starvation, Leopold began to think in systems, how ecology is an interdependent system that human behavior can really screw up, even destroy, and with that destruction lose the benefits (now called ecosystem services) that humans had assumed would always be there. By 1924, Leopold had convinced the U.S. Forest Service to set aside 500,000 acres as the Gila Wilderness Area, the National Forest System's first officially designated wilderness area (Gibbons, 2014). He left the Forest Service and spent his influential career at the University of Wisconsin grooming his students to view the human and natural worlds as one system (Meine, 2010).

In conclusion, why have I told you about Dr. Swallow and Dr. Leopold? Because they were the launchpads for Ecosystem-based Management (EBM), the dominant desired paradigm for environmental management globally.

EBM is a powerful contemporary paradigm shift that challenges Cartesian logic to reconnect and reintegrate culture, emotion, and aesthetics with biophysical science (Senecah et al., 2006). And that is what the chapters in this book demonstrate so well: how communities, alone or in processes with government ENRM agencies, are advocating or agitating for their identities, their sense of self-in-place, to have an influential voice in determining their desired futures.

A keyword search takes you to an enthusiastic cornucopia of enthusiastic, multidisciplinary, multicultural EBM resources, literature, research, policy, and applications. EBM focuses on marine and coastal management because these include interior watersheds and thereby cover pretty much all land masses, water bodies, and human uses, no matter the political boundaries. Note EBM's holistic approach instead of Cartesian-driven, traditional resource management approaches that would have us focus on a single species, sector, activity, or concern. EBM's integrated approach seeks to manage human activities because this is the only way that ecosystems will remain healthy, productive, and resilient so that they can provide the services humans want and need, and this includes leaving nature alone (McLeod & Leslie, 2009). As I noted at the beginning of this Foreword, today's ENRM challenges are not environmental problems, they are problems with human appetites, priorities, and behaviors. EBM is science-based, but also considers the interdependent and cumulative impacts of different human sectors, including cultural, social, and economic activities.

Here are a few EBM principles that I think are especially relevant to these chapters and to future research and practice:

- Collaborative governance processes, working together with citizens, landowners, businesses, local governments, interested organizations, and others to face problems, identify opportunities, make feasible improvements, and find common solutions. The processes are often as unique as the situation, but the common theme is the active participation of partners. Hence, EBM seeks to build cooperative, long-term alliances with communities to implement EBM.
- All forms of relevant information, including scientific, indigenous, and local knowledge, innovations, and practices. Assumptions behind

proposed management decisions should be made explicit and checked against available knowledge and views of stakeholders.
- Involving all relevant sectors of society and scientific disciplines, all the necessary expertise and stakeholders at the local, national, regional, and international levels, as appropriate.

EBM should continue to develop its emphasis on and consideration of these principles, and anthropological perspectives have a key role to play in this, providing a vital close understanding of stakeholders' knowledge and views. Without this, it will be difficult to implement legitimate, just decisions and change.

The Ecosystem-based Management paradigm has emerged over the past 15 years and is increasingly officially embraced via government and agency directives, policies, and application in recognition that *sector-based management* is insufficient. If humans want to continue enjoying ecosystem services, then we need to act as responsible parts of all ecosystems. Science, policy, culture, and practice need to work together to address complex and contentious environmental/natural resource management problems (Cordell & Bergstrom, 1999; Burroughs, 2011).

The chapters of this book demonstrate the rich value gained by honoring and including communities' distinct senses of who they are in their distinct place and how they, alone or with other decision-makers, can create space to include and celebrate their culture and heritage. "The world is changing quickly and our models of learning, communicating, and acting must change accordingly" (Meffe et al., 2002). As John Allen, supervisor of Deschutes National Forest in the USA, notes, "Using an ecosystem services perspective is like moving from black and white to full-spectrum color in terms of the richness of the analysis and the ability to communicate it to the public." I expect to cross paths or collaborate with Environmental Communication and Anthropology colleagues in the significant arena of Ecosystem-based Management. We have much to offer.

Chatham, NY, USA Susan L. Senecah

References

Blumer, H. (1986). *Symbolic interactionism: Perspective and method.* University of California.

Burroughs, R. (2011). *Coastal governance.* Washington, DC: Island Press.

Cantrill, J. G., & Senecah, S. L. (2001). Using the "sense of self-in-place" construct in the context of environmental policy-making and landscape planning. *Environmental Science & Policy*, *4*(4–5), 185–203.

Clarke, R. (1973). *Ellen Swallow: The woman who founded ecology.* Chicago: Follett.

Cordell, K. H., & Bergstrom, J. C. (Eds.). (1999). *Integrating social sciences with ecosystem management.* Champagne: Sagamore Publishing.

Dyball, R., & Carlsson, L. (2017). Ellen Swallow Richards: Mother of human ecology? *Human Ecology Review*, *23*(2), 17–28.

Environmental Communication Division. (n.d.). National Communication Association. Retrieved February 12, 2021 from https://sites.google.com/view/ecdnca/home/

Gibbins, T. (2014, Summer/Fall). An American original: Aldo Leopold and the Gila Wilderness. *Your National Forests Magazine*, (Summer/Fall), 22–24.

International Environmental Communication Association. Retrieved February 12, 2021 from http://theieca.org/

Jacobs, J., & Batten, J. D. (1967). The three little pigs. *English fairy tales.* New York: Dover Publications.

Lawrence, R. J. (2001). Human ecology. In M. K. Tolba (Ed.), *Our fragile world: Challenges and opportunities for sustainable development* (pp. 675–693). Oxford: Encyclopedia of Life Support Systems.

Leonard, K. (2020, December 20). *Brother ocean, sister lake: Why bodies of water deserve respect and human rights.* Let's Talk About Water. Retrieved February 11, 2021 from https://www.letstalkaboutwater.ca/post/brother-ocean-sister-lake-why-bodies-of-water-deserve-respect-and-human-rights/

Leopold, A. (1969). *A Sand County almanac.* New York: Oxford University Press.

Lyman, H., Lyman, S. M., & Blumer, A. J. V. (Eds.). (2001). *Selected works of Herbert Blumer: A public philosophy for mass society.* Champaign: University of Illinois Press.

McLeod, K., & Leslie, H. (2009). *Ecosystem-based management for the ocean.* Washington, DC: Island Press.

Meffe, G. K., Nielsen, L. A., Knight, R. L., & Schenborn, D. A. (2002). *Ecosystem management: Adaptive, community-based conservation.* Washington, DC: Island Press.

Meine, C. (2010). *Aldo Leopold: His life and work.* Madison: University of Wisconsin.

Nagourney, A. (2007, February 25). Gore Wins Hollywood in a Landslide. *The Caucus. The New York Times.* https://thecaucus.blogs.nytimes.com/2007/02/25/gore-wins-hollywood-in-a-landslide/

National Communication Association. (2021, February 10). *What is communication*. https://www.natcom.org/about-nca/what-communication
New Zealand Parliament. (2017, March). *Innovative bill protects Whanganui River with legal personhood*. Retrieved February 11, 2021 from https://www.parliament.nz/en/get-involved/features/innovative-bill-protects-whanganui-river-with-legal-personhood/
Samson, F. B, & Knopf, F. L. (2001). Archaic agencies, muddled missions, and conservation in the 21st century. *BioScience*, *51*(10), 869–873.
Senecah, S. L, Manno, J., Wise, W., & Conover, D. (2006, June). *Ecosystem-based management in New York State: Taking the next steps. A summary of contributions by participants at five statewide dialogues in 2006*. New York State Department of State. https://www.oglecc.ny.gov/media/EBM%20Dialogues%20Report%20FINAL.PDF
Senecah, S. L. (2004). The trinity of voice: The role of practical theory in planning and evaluating the effectiveness of environmental participatory processes. In S. P. Depoe, J. W. Delicath, & M.-F. Aepli Elsenbeer (Eds.), *Communication and public participation in environmental decision making* (pp. 13–33). SUNY Press.
Smith, A. V. (2019, September 24). The Klamath River now has the legal rights of a person. *High Country News*, pp. 22–23.
Trivizas, E., & Oxenbury, H. (1993). *The three little wolves and the big bad pig*. New York: Margaret K. McElderry Books: Maxwell Macmillan International.
Zehr, P. E. (2021, January 27). Damn you and your dumb dualism, Descartes. *Psychology Today*. Retrieved February 8, 2021 from https://www.psychologytoday.com/us/blog/black-belt-brain/202101/damn-you-and-your-dumb-dualism-descartes/

Praise for *Anthropological Perspectives on Environmental Communication*

"Adaptive and inclusive governance, key aspects of steering towards sustainable use of land and water, require cross-cultural and respectful communication. A basic prerequisite is insight into the diversity of ways that people perceive and interact with the rest of nature. This book brings to the fore the significant contributions that anthropological research offers in this respect. The volume not only provides a number of highly interesting and theory-framed cases of the multifaceted relationships between people and their local environments from various parts of the world, it also stimulates a healthy uncertainty regarding the context and rationale for human actions, which encourages more open-minded approaches in environmental communication and mutual exchange in the search for sustainable solutions."

—Marie Stenseke, Professor in Human Geography, Deputy Dean of the School of Business, Economics and Law, *University of Gothenburg*, and Co-chair of IPBES Multidisciplinary Expert Panel

"Nature, culture, and people are generally perceived as independent entities in the world around us. The environmental situation presents us with new challenges within the confines of our lives on Earth. Researchers seek an explanation for human societies' various strategies, some of which are rooted deeply in the past; they reveal a faith in the value of models linked to heritage in the form of rituals, social relations, and work techniques, as well as adaptations prompted by changes in the landscape and environment. This collection of studies documents the ongoing processes of diversity that are grounded in environmental and cultural differences across the globe, taking the reader from South and North America (Chile, Arizona, Alaska) to Northern and Central Europe (Sweden, Slovakia). The results demonstrate the surprising diversity of historical contexts of place versus environmental constraints, as well as that of life in depopulated regions."

—Viera Feglová, Emeritus Professor of Cultural Anthropology, *Faculty of Social and Economic Sciences, Comenius University, Bratislava*

"This work is more than a mere collection of case studies. The authors depict worlds where a significant part of life consists in the collective performance of stories about culture and environment. We all, in fact, take part in this performance, together with our fellow players, who participate in processes of knowledge making in varied ways."

—Josef Kandert, Professor of Social Anthropology, *Faculty of Humanities, Charles University, Prague*

"Environmental conflicts are increasingly heated, and the need to listen and hear diverse voices has never been greater. This timely volume uses case studies from around the world to understand environmental communication from many perspectives. It will help practitioners and theorists understand what is being said, by whom, and to what purpose. With cases including cultural landscapes, community voices, living and material heritage, toxic legacies, continuity and abandonment, this volume brings invaluable anthropological insights and reflections on the connections between language and place."

—Simone Abram, Professor in Anthropology and Co-Director of the Durham Energy Institute, *Durham University*

"This groundbreaking intervention is fresh, interdisciplinary, and internationally authored. It not only challenges taken-for-granted assumptions about environmental communication and anthropology, but also follows that challenge with a suite of heuristics for ways to deepen what is meant by the label "environmental communication." The editors have assembled a highly diverse group of authors and integrated their ideas into a superb example of transdisciplinary scholarship that offers insights for other scholars, for advocates, and for citizens. This is a volume that will immediately reward its readers, while simultaneously beckoning them to return over and over."

—Tarla Rai Peterson, Professor of Communication, *College of Liberal Arts, University of Texas, El Paso*

Contents

Notes on Contributors

Ján Aláč followed his studies in environmental ecology with a specialization in ecomuseology by becoming a doctoral student of anthropology at the Faculty of Humanities of Charles University in Prague. In his dissertation he deals with the one-child system of reproduction and its influence on the development and sustainability of local cultures. Since 2011 he has been the head of the Department of Documentation and Administration of Collections and the Gemer-Malohont Museum in Rimavská Sobota (Slovakia). He is the principal investigator of the applied anthropological project Rescue and Presentation of Novohrad Folk Tombstones in Horný Tisovník. In 2014, as part of the Slovak Republic's Landscape Award, the activities of this project received special recognition for rediscovering the historical memory, identity, and symbols of a place in the country by reading the legacy of the past.

Michael E. Dove is a lecturer at the Department of English and American Studies, Matej Bel University in Banská Bystrica, Slovakia. His teaching and research activities focus on translation studies, American studies, and foreign-language teaching. As a lifelong reader and admirer of anthropology, he is interested in the field's potential to contribute to other disciplines. Dove holds a master's degree in translation and interpreting.

Cristian Alarcón Ferrari is a researcher at the Division of Rural Development at the Swedish University of Agricultural Sciences in Uppsala and former Professor of Law and Director at the Universidad Academia de

Humanismo Cristiano School of Law in Chile. Ferrari holds a PhD in Environmental Communication from the Swedish University of Agricultural Sciences, and in 2017–2018 he was a postdoctoral researcher at the Department of Development Sociology at Cornell University in the US. In addition to his academic work, Ferrari has worked as a lawyer in Chile, where he has advised and represented trade unions and workers for many years.

Jan Horský studied history and in 2010 became Associate Professor of Social and Cultural Anthropology. He has been lecturing at the Faculty of Humanities and various other faculties of Charles University in Prague for 30 years, particularly on the (cultural) history of the early modern period with a focus on the European Reformation and the history of the natural sciences. He focuses his research on the philosophy and theory of historical sciences, especially on the question of historical sciences' possible answer to narrative critiques of historiography. His specialization in anthropology is the interaction of cultural and evolutionary biological factors. In this context, he has long devoted himself to historical demography and family history.

Yoko Kugo is a PhD candidate in Interdisciplinary Studies in the Arctic and Northern Studies Program at the University of Alaska Fairbanks (UAF). Her dissertation focuses on Yup'ik place names and stories about these places in the Iliamna Lake area, southwest Alaska. She studied Yugcetun (the Central Yup'ik language) at UAF for two years in order to become familiar with Yup'ik ways of understanding the landscape. Her interests include Alaskan history, Indigenous knowledge and arts, local knowledge, and geography.

Simon Larsson holds a doctoral degree in social anthropology. He works at the Gothenburg Research Institute, School of Business Economics and Law, at the University of Gothenburg. His research has a broad empirical focus but relates to a theoretical interest in governance and power in historical and contemporary societies.

Ivan Murin is Associate Professor of Social and Cultural Anthropology at the Department of Social Studies and Ethnology at Matej Bel University, Slovakia, and visiting scholar at Charles University in Prague, Czech Republic. Murin is a field ethnographer with a scientific interest in intergenerational transmission of culture, cultural heredity and heritage studies, collective memory, and symbolic communication analyses.

He coordinates research teams studying sustainability, vitality, adaptation, and microevolution processes in local societies.

Katrina Rønningen is a geographer working on environment and climate, agriculture, and natural and rural resources and policies. She holds a Dr. polit. in geography. Keywords for her environmental research include commodification and marginalization processes in rural areas, conservation and land-use questions linked to aquaculture, large carnivores, wind energy, land use and rights, and land-use pressures in indigenous reindeer herding areas. Rønningen works as a senior researcher at Ruralis – Institute for Rural and Regional Research in Trondheim, Norway.

Camilla Sandström is Professor of Political Science at Umeå University, Sweden. Her research focuses on environmental politics and the governance and management of natural resources, in particular various forms of collaborative governance and comparative analysis of policy mixes addressing natural resources and implementation processes at local, national, and international levels. Her research covers policy areas such as forests, nature conservation, wildlife, and climate change.

Susan L. Senecah is a recognized leader in creating the field of Environmental Communication. She is Environmental Studies professor emerita at the State University of New York College of Environmental Science and Forestry, where she created and coordinated the Program in Environmental Communication and Participatory Processes. She calls herself a "pracademic" to reflect the intersections among her teaching, research, and practice in process design and skills for stakeholder engagement, collaborative decision making, and conflict resolution, and as Special Assistant for Environmental Policy in the New York State Senate. She holds a PhD in Communication Studies at the University of Minnesota.

Annelie Sjölander-Lindqvist is Associate Professor of Social Anthropology and Senior Lecturer in Human Ecology at the Gothenburg Research Institute and School of Global Studies, University of Gothenburg, Sweden. Her research addresses problems of sustainable development and conflicting values in areas of contested natural resources and landscape management, with attention to the limits and possibilities of participation in the context of power dynamics involving state authorities and local populations, local concerns in relation to policy decisions, and the analysis of the conjunction of local identity, science, and politics in environmental contests, considering the link between place attachment, landscape dynamics, and local involvement.

Richard Stoffle is a professor in the School of Anthropology, University of Arizona, Tucson, Arizona. He is the principal investigator for a number of research studies funded by various federal agencies, such as the National Park Service, who funded the Ethnographic Overview and Assessments at Arches National Park, Canyonlands National Park, Hovenweep National Monument, and Natural Bridges National Monument. His research team facilitates the sharing of important cultural heritage understandings between museums, interpretative programs and Native Americans in the United States and African-Ancestry people in the Caribbean. These heritage studies potentially contribute to more accurate understanding of the past, increased sensitivity in policies, and better public education.

Kathleen Van Vlack is an applied anthropologist who co-founded the non-profit applied anthropology firm Living Heritage Research Council. She serves as the non-profit's Vice President of Research. She has participated in a range of ethnographic studies, such as environmental impact assessments and cultural heritage preservation studies across the United States, and has worked with over 50 Native American tribes. Van Vlack has won a number of awards and research grants, including the Ruth Landes Memorial Research Fund for Dissertation Research and the Friedl and Martha Lang Student Award from the High Plains Society for Applied Anthropology.

List of Figures

Dancing with Lava: Indigenous Interactions with an Active Volcano in Arizona

Arsenic Fields: Community Understandings of Risk, Place, and Landscape

Cultural Transmission in Slovak Mountain Regions: Local Knowledge as Symbolic Argumentation

Community Voices, Practices, and Memories in Environmental Communication: Iliamna Lake Yup'ik Place Names, Alaska

Demographic Change and Local Community Sustainability: Heritagization of Land Abandonment Symbols

Living Stone Bridges: Epistemological Divides in Heritage Environmental Communication

Power, Conflicts, and Environmental Communication in the Struggles for Water Justice in Rural Chile: Insights from the Epistemologies of the South and the Anthropology of Power

List of Tables

Introduction

Annelie Sjölander-Lindqvist

This book is a collection of chapters that explores and discusses the influential and pervasive role of communication on issues pertaining to the environment. Every day we read, hear, or see communication about environmental issues: in regular news media, on Facebook, through tweets, on TV, at work, at school, or at parties. While there are countertendencies and a movement that denies climate change and downplays environmental issues, the issues of climate change, loss of biodiversity, and overexploitation of natural resources continue to make up a large proportional part of the societal debate. Issues related to the environment are truly global and influence people all over the world in the contemporary epoch of the Anthropocene—but both consequences and responses to environmental crisis differ depending on local politics and local dependency on the environment, as well as cultural understandings.

This volume of chapters was initially presented at the 2018 meeting of the Society for Applied Anthropology in Philadelphia, Pennsylvania,

A. Sjölander-Lindqvist (✉)
School of Global Studies & Gothenburg Research Institute, University of Gothenburg, Göteborg, Sweden
e-mail: annelie.sjolander-lindqvist@gu.se

A. Sjölander-Lindqvist et al. (eds.), *Anthropological Perspectives on Environmental Communication*, Palgrave Studies in Anthropology of Sustainability, https://doi.org/10.1007/978-3-030-78040-1_1

USA. The session "Anthropological Contribution to Environmental Communication" grew from the concern about sustainability and the wish to explore the different ways individuals and collectives experience, understand, and act upon environmental challenges. The contributions to this volume proceed from the premise that anthropology is well positioned to contribute to the study of environmental communication.

Anthropology's focus on the holistic dimensions of the human condition, its interest in understanding humankind's cultural variation wherever it occurs, and its sensitivity to both similarities and differences, while never losing sight of the powers of politics, ideologies, economies, and ecologies, provide a solid foundation for such a contribution. This book tries to make sense of how an anthropological perspective can further our understanding of the diversity of environmental communication and the different ways people—verbally and non-verbally—communicate about and with their surrounding environments. An important aspect lies in the discussion of the communicative prospects for sustainability, seeking to bring anthropology into more theoretically and empirically productive engagement with the study of environmental communication.

Environmental Communication

Within the overall argument for much-needed transformative environmental change (Díaz et al., 2019; Sygna et al., 2013; United Nations, 2015)—be it radical through deep systematic and structural shifts that challenge our assumptions, beliefs, and values (Armitage et al., 2017) or gradual through changes to the existing order rather than a radical rebuilding of it (Armstrong, 2006)—communication remains a crucial and critical element. This is reflected, for example, in the utilization of communication as a strategy to inspire people to behave in ways less destructive to the environment (e.g. McAfee et al., 2019) and the prevalent employment of collaborative- and partnership-directed governance processes to spur new insights and ideas, and to increase political legitimacy (e.g. Chaffe et al., 2016; Sandström et al., 2018; Valadez, 2018). Other examples are provided by the role of artistic media and games in spurring awareness of environmental concerns and swaying public opinion (Brady, 2011; Fjællingsdal & Klöckner, 2020; Morrison, 2018). The study of negotiations and struggles surrounding the construction of biogas plants and other ecological modernization projects and plans is another field of interest with relevance for environmental communication, where the intent is to discern the role of intersubjective communication in issues pertaining to the environment (e.g. Alarcón, 2015; Alarcón, chapter "Power, Conflicts, and Environmental Communication in the Struggles for Water Justice in Rural Chile: Insights from the Epistemologies of the South and the Anthropology of Power" of this book; Walker et al., 2019). Descriptively speaking, environmental communication can be said to constitute a field of practice, research, and scholarship in the nexus of environment and human communication that "can play an important role [...] in understanding, critically analyzing and facilitating transformations to more sustainable and just societies" (Joosse et al., 2020, p. 10).

Recent years have witnessed a growing interest in the role and societal value of environmental communication research, in particular as scholarship finds that communication can enhance policy implementation (e.g. Senecah, 2004; Zikargae, 2018). Whereas the early years of environmental communication scholarship largely focused on the rhetorical aspects and the concrete practices of communication (e.g. media agenda-setting), the scope has since widened (Cox & Depoe, 2015). Today, the study and practice of environmental communication are more diverse, including but not restricted to community engagement in shared-resource governance, environmental journalism, advocacy campaigns, science communication, risk communication, environmental justice movements, social media messaging, and any other practice where there is a flow of information, unidirectional or interactive listening, public discussion, and debate (Evans Comfort & Park, 2018; Cox, 2013; Hansen & Cox, 2015; Zikargae, 2018). Anders Hansen and Robert Cox (2015, p. 8) argue that the field of environmental communication has consolidated itself "as a distinctive subfield of media and communication research" and advanced the "understanding of the complex processes involved in the social 'construction' of the environment as an issue for public and political concern." As of today, climate change communication, sustainability science, visual communication, and the problematizing of the human–nature binary stand at the forefront of environmental communication scholarship (Cox & Depoe, 2015).

Anthropological Perspectives on Environmental Communication

Since the early days of anthropology, anthropologists have been interested in human diversity and real-world contexts, what happens between people in certain settings, what drives people's understanding of the surrounding world, and how people act/interact with the environmental surroundings of their homes. Communication stands at the heart of this, or, as proposed by Gregory Bateson in *Steps to an Ecology of Mind* (1972): all kinds of organization are by nature "communicational." Communication can be the sharing of information or a way to convince of or propagate certain actions, but it can also be what Edmund Leach (1976) referred to as "*all* the various non-verbal dimensions of culture" (p. 10). Styles in clothing, ways of eating and cooking, architecture, place names, myths, cosmologies, knowledge and traditions, and body gestures are all reflections of "coded information in a manner analogous to the sounds and words and sentences of a natural language" (Leach, 1976, p. 10).

When information, ideas, and knowledge are transmitted and disseminated, the meanings reach us, the receivers, through symbols and signs, in written, oral, visual, and/or sensorial forms (Turner, 1977). This communication can be clear to the observer, but it can also be indirect and subtle, in disguise. Regardless of the form it takes, the sending and receiving of messages is perceptual, based on recognition and present in almost every moment of our lives. Someone speaks, someone listens, someone acts, someone observes. It is a relational activity that engages our senses, situated and manifested through symbols and behavior. When we move through life and do "our things," we not only talk, see, and hear, but we use all the senses in our communicative acts. When we eat, we use our senses of taste and smell; when we search for remedy for and relief from an illness, touching is an important medium for locating and explaining what may be wrong; when we dance, we reflect concepts of body and space; and when we liven up a digital text message with an emoji, we wish to communicate feelings such as joy, disappointment, anger, or happiness. Likewise, when we drive a car, we signal with our blinkers that we are turning right, and when we put up a warning triangle, we tell fellow road users to be alert. During these moments, we establish contact and convey information using various media, and in doing so we exchange views, beliefs, and assumptions that can be explicit or implicit.

Communication is in this sense both imagination and performance, and is not only about the sharing of information. As a relational act, it is also a process where we share, hope that we are being heard, and hope that what we say, intentionally and unintentionally, shall be known, felt, and experienced. What is said, done, and interpreted are all constituent parts of the communication process, whether it is face-to-face discussions at public meetings, political graffiti, alerting risk and danger through information campaigns, or the naming of places to signal proper resource use and importance to local livelihoods. As suggested by Eugene Anderson (2014), we should not only look at what people think, say, and share when it comes to nature and environment, but also acknowledge the interactive context as marking the beginning and the end to the way we imagine ourselves and our place in the world (cf. Abram, 1996).

Buying a detergent labeled as environmentally friendly may be a tacit way of saying "I wish to contribute to sustainability," and that we, consumers, have faith in the power of everyday life to bring about change in the future (Pink, 2012); when we consume a beer with a logo depicting a local mountain, the brewer speaks to us about a cherished place (Sjölander-Lindqvist et al., 2020a). Others have shown how the ambition to develop roads and other transportation networks, seeking to deliver social integration, economic development, and increased modernization, may paradoxically lead to more deaths; the slow road that has turned into a highway may also become a route punctuated by memorial shrines to commemorate those who have lost their lives to motorized collisions (Harvey & Know, 2015). A toxic spill may become a subject of communication among community members and lead to a recognition of environmental hazards. The trees, the community ponds, and the small brooks in people's environs serve to tell them about the lives of past generations and the collective and inherited environment (Sjölander-Lindqvist, 2004). Yet other anthropologists have shown how environmental movements who lay claim to authority over landscapes and biodiversity resources essentialize certain values over others, leading to the silencing of different groups who may be dependent on the resources of the forests, the waters, and the agricultural lands for their livelihoods (Heatherington, 2010). Communication may therefore be comforting, contradictory, disturbing, or supportive. It can be a vehicle for positing actions, strategies, policies, and the messages embedded in ideological and worldview-shaped conceptual frames (cf. Lakoff and Johnson, 2003 [1980]; Underhill, 2011), representing our concerns about the uncertain and the unstable (Sjölander-Lindqvist et al., 2020b).

In his study of climate perceptions in the Peruvian Andes, anthropologist Karsten Paerregaard (2020) finds that "climate communication is a diverse and complex enterprise embedded in a web of social relations and cultural interactions that transform its message and ascribe it new meanings" (p. 123). Werner Krauss and Hans von Storch (2012) suggest that overcoming the discrepancy between the global climate model and the fact that people do not always see climate change as a personal threat or even relevant to them requires a broader knowledge approach (cf. Brevini, 2016). Paerregaard continues, "even though climate change is a global phenomenon, people experience it locally often as part of other processes of change" (p. 123). If climate change is experienced as psychologically distant (Spence et al., 2012), or if the presence of toxics in the immediate local environment is a disregarded risk (Sjölander-Lindqvist, chapter "Arsenic Fields: Community Understandings of Risk, Place, and Landscape" of this book), it is not because people are ill informed or "irrational" (Spence et al., 2012). Failure may instead lie in the fact that communication about environmental change, risks, and perils has failed in channeling attention in ways that "speak to people" and their concerns (Sjölander-Lindqvist, chapter "Arsenic Fields: Community Understandings of Risk, Place, and Landscape").

The difficulty of explaining the environmental hazards of climate change to the public is, as suggested by Cox and Depoe (2015), a topic requiring greater concern. Traditional, single-media-focused communication approaches are less efficient in inspiring the individual to behave in a more climate-friendly way, which poses a challenge for the policy sector to stir up communication to increase awareness about the causes and consequences of climate change (Moser, 2016; Semanza et al., 2008). However, as argued by Cristian Alarcón (chapter "Power, Conflicts, and Environmental Communication in the Struggles for Water Justice in Rural Chile: Insights from the Epistemologies of the South and the Anthropology of Power" of this book), scholarship should strive to address these questions while being aware of normative priorities, since environmental communication strategies are often accompanied by the implementation of norms on, for example, governance to ensure sustainability (Sjölander-Lindqvist et al., 2020b). The Rio Declaration on Environment and Development, for example, and Agenda 21 refer to democratic decentralization as a key component of good governance. Greater public engagement through consultation, negotiation, and cooperation in policy design and implementation can generate a more heterogeneous pool of knowledge, which in turn can improve the quality of decisions (Primmer & Kyllönen, 2006).

However, including different ways of knowing requires the involved parties to deal with epistemological as well as practical aspects of relating to different knowledge spheres (Risvoll & Kaarhus, 2020; Sjölander-Lindqvist et al., 2020b). Alarcón cautions us to be aware of how knowledge and power give rise to what he refers to as "communicative and epistemological struggles," phrased by Andréa Zhouri (2018) as "epistemological violence": when local and traditional ways of understanding and being in the world are dismissed in consequence of ecological modernization. In line with the debate on ethnographic representations (Clifford & Marcus, 1986; James et al., 1997; Katz, 1992) of the "distinctive manners of imagining the real" (Geertz, 1983; p. 184), both Richard Stoffle and Kathleen Van Vlack, in their contributions to this book, raise the point that we, as scholars, must ask who is speaking and who is considered an expert or authority to avoid the reproduction of discursive colonialism (cf. Kugo, chapter "Community Voices, Practices, and Memories in Environmental Communication: Iliamna Lake Yup'ik Place Names, Alaska" of this book; cf. Wassef, 2001). These thoughts call into question the subject positions associated with power, challenging a utilitarian and normative understanding of environmental communication in the fields of climate change, biodiversity loss, toxic contamination, overuse of natural resources, or any other field where the environment is at stake. Recognizing the physicality of environmental change is not enough: the social, economic, cultural, and political dimensions of environmental management, the positionalities of different actors, and the epistemic status of their ways of knowing must be accounted for.

Mapping the Contribution

Central to the anthropologically informed inquiry, then, is how we make sense of our world and everyday realities, be it through the eyes of an institutional or residential individual or collective, in the context of climate change, loss of biodiversity or wildlife due to overuse, or changed prospects for livelihoods in consequence of environmental conservation politics. From the outset of our work on forestry, coastal management, fishery, and land use in Europe, North America, and South America, the authors of this volume are united in the understanding that environmental communication spans both the unique and the conventional, is broad in scope, and includes different ways to communicate environmental issues. As the different cases discussed in this book demonstrate, the ways people make sense of shared spaces and of their experiences and knowledge of living in a certain place can enrich the understanding of environmental communication as a practice and process, framing and conveying intentions, reasons, and arguments for establishing awareness and readiness for action in a changing world. This relates to the issue of the way we understand and represent the environment, and how we have adapted to, and continuously make sense of, environmental circumstances.

This position is directly related to theoretical debates in environmental anthropology concerning the role of the discipline in contributing to just solutions and the imagination of a better future (Kopnina & Shoreman-Ouimet, 2013; Kottak, 2010). We know that through formal education, so prevalent in the world today, we can acquire a highly equipped mind and the ability to think analytically and make pragmatic decisions. Still, environmental challenges remain, and though sustainable development has long been on the agenda, we are far from achieving sustainability. Such challenges call for an attentiveness to the particular and the micro level, awareness of other, more macro levels, and a comparative focus to understand variations and similarities in the cultural signification of existence and experience. In this endeavor, we move toward understanding the myriad ways of life and a more expansive understanding of environmental communication. Seeing communication as an intrinsic part of the human condition and environment as a culturally specific context and the product of particular historical and cultural configurations (Bateson, 1972; Fitzgerald, 1993; Leach, 1976; Titsworth et al., 2021) lends the perspective that any arena and setting provides a window into different spoken and unspoken registers of meaning. These meanings emerge from human engagement with the landscape and reflect understandings of human existence and human society (Ingold, 2000; Rival, 2001), and come into being through concerted, fortuitous, embodied, and lived practices. What the contributions to this book have to say is that we need to acknowledge the experiential, deep-rooted, and symbolic meanings people hold.

Peter Jordan's (2003) study of Siberian Khanty hunter-fisher-gatherers provides an ethnographic portrait of how local communities are engaged in what he calls "dialogues of place" when people interact with places in the landscape. "Life is a state of dialogue" (p. 281), taking place symbolically and materially at both individual and collective levels, Jordan argues. Every animal and fish is part of the landscape, and individuals need to interact with these animals and take care of them and the landscape to maintain the totality of life. The islands, the waters, the rapids, and the high grounds are the venues and the locales for "a communicative relationship of obligation and reciprocity," which is a basic condition for the welfare of Khanty communities. The Khanty community is just one of many examples of how generations of people have contributed to creating the ways our cultures interact with and adapt to environments on a trial-and-error basis (Mesoudi et al., 2006). The exploration of how communities ideographically record environmental phenomena and keep their experiences and knowledges in the collective consciousness proves important in learning about the causality of temporal and spatial relationships, often in a complex chain of meaningfully interconnected, yet discrete things.

More than many other conventional disciplines, anthropology and its interest in and concern for humankind and the everyday, as it unfolds in different settings and contexts, can reveal more of the inner world of both the generations who have long been in immediate contact with the organically changing environment and all those other actors involved in imagining the everyday and the future. How people modify, symbolize, and adapt to their immediate surroundings has been a central part of anthropology since early on. What we do and why are questions whose answers lie in ideas, aspirations, norms, and values, sometimes shared, at other times disputed. These are all parts and dimensions of what creates and reaffirms life, as well as the creativity involved in developing ways of living. The way that people learn about their environment and develop elaborate co-adaptations with it is discussed in the book *Man's Role in Changing the Face of the Earth* (Thomas Jr. et al., 1956). Its chapters present various academic findings that suggest humans can and do influence natural processes and are not just passive components of the Earth's ecosystem. Omer Stewart contributed a chapter based on his fieldwork, arguing that Native Americans used fire to shape their ecosystems in the High Plains of North America (1956) (cf. Stewart et al., 2002). This debate continues today (James & Marcus, 2006; Stewart et al., 2002), as demonstrated, for example, in Minh Nguyen's study (2016) of migrant waste traders who regenerate and revalue urban space through pioneering local trade in recyclable waste as a means to earn an income and sustain their livelihoods. This points to the importance of engaging with local matters, stressing close ethnographic attention to the everydayness of the individual and the collective—from households, neighborhoods, and villages and to all other spaces where human experience and consciousness unfold, in any form it may be represented and made sense of.

Anthropology has long championed "the other's view of the world." Beginning as the study of non-Western small-scale societies, anthropology acknowledges cross-cultural understanding through focusing on a plurality of voices to offer nuanced perspectives and knowledge regarding the ways people around the world see and think of the world, what they say about their lives, and why they do what they do. Åsa Boholm (2015) defines this task as reconciling emic and etic perspectives. This is not only central to anthropological scientific inquiry; it is perhaps even more important in a world periled by environmental and social challenges, where there is a risk that practical executions of environmental-protection visions may lead to people and their concerns, needs, and knowledge (continuously) being (even more) marginalized (e.g. Kellert et al., 2000; Lam et al., 2020; Sjölander-Lindqvist, 2008, 2009; Sjölander-Lindqvist et al., 2020b). While just, equitable, and sustainable transformation is a significant element in policy and politics, realizing these goals is another matter. In a time of changing climate and scarcity we might be more likely to see exacerbated conflicts and inequality in terms of power distribution, social, and economic gains (Blythe et al., 2018; Lam et al., 2020).

Embracing the diversity of values is fundamental to achieving societal goals for sustainability (Pascual et al., 2014). The prospects for realizing crucial and just change lie in an inclusive incorporation of different values—be they, for example, instrumental or utilitarian values (to achieve human ends), intrinsic/ethical values (inherent to nature), or a combination of the two. Comprehending the depth of the perception of the environment held by the person "across the table" requires cross-cultural understanding. Regardless of the point of departure or focus, communication is a crucial tool for channeling attention to environmental perils and the communicative aspects of change and existence. Here lies the importance of acknowledging how different groups speak about and understand the environment, and how to represent this to others.

This reflects the need for cross-cultural understanding and the role of language. Overly technical jargon or an unfamiliar native language or dialect can lead to misinterpretation, which can in turn prevent the message of the communication from being turned into the action that the sender wanted (Boholm, 2015). Considering that language is often designed to reduce, filter, and control the communication of ideas (e.g. Sjölander-Lindqvist et al., 2020b), it can, together with unconscious and conscious bias, create barriers, as language is also the performance of identity (Bassiouney, 2018), while a dialect can be a way to resist power. Take, for example, how conversational practice and dialectal difference in Mesoamerica encode and mark colonial resistance, revealing an unbreakable link between traditional culture, identity, and present-day life (Romero, 2015). Therefore, while an approach for change can be designed with the best intentions, it may not be truly transformative for the targeted groups and communities due to neocolonialism. This was the case in Mozambique, for example, where cultural and historical power dynamics have impacted health communication (MacLeod & MacDonald, 2018). Historically rooted friction and suspicion can lead language to reduce the contents of the mind, leaving communication incomplete and ambiguous (Dávid-Barrett & Dunbar, 2016). To reveal other mental worlds, distinct cultural configurations, and different taxonomies, we need to direct our attention to the importance of environmental issues as a source of collective representations of local mental worlds and the immanent values held in relation to the environment. Here, an anthropological approach to communication can complement the field of environmental communication, shedding light on the need for us to realize that the loss of local culture and environment changes the consistency of values and actions. The anthropological gaze provides insight into which impacts are significant to culture and environment.

Over the years, the contributors to this volume have met and conversed with local residents, community members, agency officials, NGO representatives, foresters, farmers, and fishermen. And the list of all who have contributed their time and effort to make our research possible does not, of course, end here. Our interest in thought and action, in the shared and the disputed, echoes plural meanings and tacitly held assumptions about existence and society. Our interest in the observation, interpretation, and analysis of relationships, experience, conceptual structures, ideas, assumptions, aspirations, and values can be summarized as striving to understand what makes life. Another way of describing this thesis is that anthropologists, by collecting information which may be sampled using various methods, "pin down facts about people" (Sobo & de Munck, 1998, p. 16) in order to describe the variety of culture and society. Often, we have not only met our informants and talked to them formally and informally; some of us have engaged in workshops and focus groups, and we may, through the method of participant observations, have been engaged in various parts of local life. During these meetings, many of our informants have shared their feelings, concerns, and fears with us. This lies at the center of anthropology: when our informants tell us their stories about life as it unfolds in their own social and organizational setting—their family, household, village, neighborhood, or their wider area—we are offered glimpses of their life worlds and lived realities.

When shown a hatch to a rustic cellar, we may be told about how the small and empty well in the rather cramped room used to supply fresh potable water not only to one family, but also to two additional family farms during dry summers. These stories—as well as the many that we share with you in this book—are telling examples of the meanings we attribute to our surroundings over the course of life. What the story of the small well tells us is that it did not only serve as a resource for drinking water, but the well was also a symbol signifying community life, lived reality, and identity (Sjölander-Lindqvist, 2004). Everyone can easily remember similar events when we, through our own actions, perceived and experienced the dynamics of our surrounding environment or culture. Culturally and socially rooted, these occasions unfold and conventionalize beliefs, values, norms, and knowledge. These moments define and figurate meaning, making the exchange of experience, memories, and knowledge actionable and applicable.

It can be expected that the assessment of the experience of human cultures, that is, the cognitive, behavioral, embodied, and material outcome of humanity's previous interactions with the environment, will with equal significance affect the attitudes we adopt in the future (Morin, 2016). As an example of this, if we accept the value of stone arches only as rock formations and subject to the continual degradation of natural processes, we will adopt reductive narratives in environmental communication (Stoffle, chapter "Living Stone Bridges: Epistemological Divides in Heritage Environmental Communication" of this book). However, if we accept the role of the stone arch as a cultural representation of life, experience, and knowledge, hiding behind the horizon, changing, shaping, and offering messages, shade, or shelter, we also accept that our forests, the trees, the waters, the mountains, and the lands surrounding us signify and embed experience and knowledge, thereby rendering meaning to the ways we exist and act. The anthropological view of environmental communication is that is contingent and embedded, driven by contextual probabilities and attributes.

This creates what Toda (1976) refers to as a "nested" situation. This "nestedness" is a way of formulating the core of anthropological exploration; situations and activities where interaction among both human and non-human entities will be confined by socially and culturally framed tangibles and intangibles, emotions, and value-driven circumstances. As phrased by anthropologist Victoria Strang (2004), any situation is "the result of specific social, spatial, economic and political arrangements, cosmological and religious beliefs, knowledges and material culture, as well as ecological constraints and opportunities" (p. 5, cf. Brondizio et al., 2009). The nested character of life requires that we not only translate between languages, but also tolerate, acknowledge, and appreciate social and cultural variation and interpersonal interaction to better understand what drives our actions in the world, be it lived as a farmer, a hunter, a manager, an academic, or all of them combined. These are all contexts of learning that are provided to us when we approach and encounter different settings and meet different people, who all have their own unique experiences and knowledge, and are situational, located, and far from conforming to one another. As described by Arturo Escobar (2001), "place continues to be important in the lives of many people, perhaps most, if we understand by place the experience of a particular location with some measure of groundedness (however, unstable), sense of boundaries (however, permeable), and connection to everyday life, even if its identity is constructed, traversed by power, and never fixed" (p. 140).

Our focus is to add perspectives to environmental communication by utilizing insights from anthropology. This book proposes a closer look at the ordinary and the particular, going behind and beyond environmental advocacy rhetoric, emphasizing stronger anthropological engagement to develop understanding and knowledge of the co-constructive character of environmental communication practice. Our task, however, is not only to describe and compare; it is equally important for the anthropologist to explain and demonstrate how different elements are tied together. This interconnectedness is perhaps even more important to understanding the complexities of environmental communication and moving toward reinvigorated anthropological studies attending to the nuances of the verbally and non-verbally expressed and communication as basic to human life and culture. We suggest it is necessary to be conscious of the conceptual and the practical, as well as the verbal and the non-verbal, in order to understand how we, as human beings, create and express meaning (Hylland Eriksen, 1991; Ingold, 2010; Geertz, 1973; Worth & Adair, 1975).

By exploring the different ways people's voices are communicated, perceived, and manifested, be they in reference to facts, shared or disputed values, sameness and difference in terms of interests and goals, about the untold, or ignoring the possible existence of other ways of conceptualizing the world, this book recognizes and appreciates the world as seen, as experienced and known, as felt by human senses, and as presented and represented to ourselves and others. Communication may take forms verbal, non-verbal, situated, spatial, temporal, and sensorial, and a message may be one or all of these at the same time. As such, communication is a constant becoming, marking the beginning and the end to the way we imagine ourselves and our place in the world.

The Chapters

The ensuing chapters will take you to different places around the world (see world map on p. v), and we will see how anthropological theory and method can extend our knowledge about environmental communication, the different forms it takes, and what informs the communication process. These case studies demonstrate how communication, at the nexus of the environment and the human, can be about an intention to inform about policies and incentives in a collaborative- and partnership-directed governance process, and how environmental communication can also be about imagining what makes us, human beings, rooted in the world. Drawing on ethnographic methods in critical explorations, the contributors to this volume make both direct and indirect reference to the importance of the richness of details provided in the field using participant observations, in-depth interviews, informal conversations, and other field techniques. The details that the anthropologist gathers are crucial to the understanding of the contextual value-driven circumstances and the formal and informal rules, resources, and norms encountered when accessing a particular bounded cultural setting such as a local community, a governance process, or an impact assessment procedure. Under the well-established concept of "thick description" (Geertz, 1973), the anthropologist strives toward grasping what is getting said and the importance of what is said, be it through the spoken word or through the occurrence of a particular tradition, the implementation of a policy, or simply anything that occurs within the ordinary and the everyday. Whereas some refer to such ethnographic exploration as a way of describing other people's lives through observing, participating, listening, and asking (Bate, 1997; Hammersley & Atkinson,

1995; Ingold, 2008; LeCompte & Schensul, 2013), it is also a framework allowing the anthropologist to compare and critically contribute to the understanding of the human condition, being in the world, and knowing the world (Ingold, 2008; Schensul & LeCompte, 2016).

Now for a closer look at the case studies themselves. In the chapter "Dancing with Lava: Indigenous Interactions with an Active Volcano in Arizona", Kathleen Van Vlack brings us to Southwest of the United States, specifically to northern Arizona and the Little Springs Lava Flow, which is a remnant of an active volcano. In her chapter, Van Vlack explores the issues of how different groups make sense of the active volcano and their ensuing different understandings of place and landscape—from the perspective of archeologists who have visited the area to record data on the volcanic eruption as a means to understand American Indian lifestyles and behavioral responses that occurred before, during, and after the event, to the Kaibab Band of Paiute Indians' interpretation of the lava flow and volcanic fields' significance in relation to the ways they have interacted with the landscape. By situating the case study in a historical context, highlighting how geology has influenced the landscape and which features have important ceremonial meaning to the Southern Paiutes, Van Vlack critically reflects on the role of knowledge and who is given voice to explain the past. This epistemology-influenced debate on whether the volcano is a risk to humans (as argued by the scientific community) or a landscape intimately connected to American Indian history and their cultural heritage (as argued by the Paiutes) lays important groundwork for the discussion of how ideology, ways of knowing, and worldview-shaped concepts lay claims to authority over representation and over landscapes themselves.

In the chapter "Arsenic Fields: Community Understandings of Risk, Place, and Landscape", Annelie Sjölander-Lindqvist discusses similar issues when she brings the reader to central Sweden in Northern Europe and explores how a contaminated community makes sense of living on or very near to arsenic fields. This chapter also attests to the role of the state in place and identity, as the plans for soil remediation, situated in the context of environmental and public health, establish a fluid zone of interpretation and reflection on collective identity, the role of collective memory in instilling a sense of community, and the sense of self in an ever-changing world. Similar to Van Vlack's study, Sjölander-Lindqvist finds the importance of addressing environmental communication from the perspective of temporality, as meaning evolves over time and by means of shared experiences, through which meaning is not only created and known, but also instilled with symbolic power. These various arrangements, or frameworks of meaning and rationales of action, become mobilized and contested in everyday life.

The chapter "Cultural Transmission in Slovak Mountain Regions: Local Knowledge as Symbolic Argumentation" takes us to Central Europe and the mountain regions of south-central Slovakia, where Ivan Murin provides us with a detailed case study on the challenges of adaptation when a new generation returns to a land their ancestors had cultivated and then left due to demographic change and state intervention. Murin's case study is not only a reflection on the consequences of modernist development, he also presents an exploratory model for how people returning to their ancestral roots can re-learn and re-connect to their heritage, entering into a process in which they can reverse the interrupted transmission of knowledge and instead re-adapt. This case of cultural transmission is also a perceptive insight into one of anthropology's core subjects, that of intercultural communication, but with a particular focus on intergenerational communication and how anthropologists can facilitate connection to a place lost, but cherished and vividly present in the collective memory of the returning generation.

In the chapter "Community Voices, Practices, and Memories in Environmental Communication: Iliamna Lake Yup'ik Place Names, Alaska", Yoko Kugo also takes on this role of facilitating intergenerational communication in her study of how place names are a form of communication between people and landscape, reflecting not only the geographical features of land but also relating to individual and collective memories. As in Murin's preceding chapter, Kugo's study is about heritage preservation and how anthropologists can be communicative supporters who help community members reflect on the meanings of their surrounding landscape. Kugo shows us how certain Indigenous place names can convey environmental and spatial information, but also emphasize the temporal and spiritual relationships between the people and the land. Kugo's study also illustrates how active engagement and participation in daily life are core to the centrality of listening and asking about the visual, the directional, and the historical, both for the anthropologist but also, most importantly, for those who have lived, and live, in the landscape. This is an example of another kind of environmental communication, a form of communication in which place names provides the local community with a route to maintaining a healthy reciprocal relationship between the human and animal worlds, as well as between the living and spiritual worlds.

In the chapter "Demographic Change and Local Community Sustainability: Heritagization of Land Abandonment Symbols", we return to south-central Slovakia and how the future of cultural landscapes is linked to current global challenges. In this case study, Ivan Murin, Jan Horský, and Ján Aláč discuss how an abandoned landscape is also an environmental problem, which, as we know, is a long-standing trend around the world. The authors rightfully ask: What happens when social learning and cultural transmission of knowledge, the sharing and acquisition of experiences and local and traditional knowledge, are put to a stop due to political, economic, and demographic change? In some Slovak regions, one such immense phenomenon was depopulation and the replacement of family farms with large-scale state-run agricultural production units. As argued by the authors, although the new generations made short-term economic progress, this arrangement proved unsustainable in the end. In the next generations, the virtue of communal sharing of the land and the associated and essential dimension of sociability were lost. The authors bring us to the cemetery to discuss how this lost heritage can be re-created through the restoration of tombstones, as these carry signs and symbols that can support the remembrance of life, death, and what was important to the local farming community.

After this second European stopover, we return in the chapter "Living Stone Bridges: Epistemological Divides in Heritage Environmental Communication" to the United States and the Natural Bridges National Monument, a protected area in southern Utah. Here, Richard Stoffle shows that environmental communication is complex and fraught by epistemological divides. Discussing the case of massive stone bridges and the diverging understandings of what makes up the world and the purpose of a particular landscape feature, he finds it opportune to address the phenomenology of landscape. This is particularly important in cases where there are no material resources or artifacts to link the discussion to anything other than the natural landscape. This makes environmental communication an intriguingly difficult project, as the parties need to turn to phenomenology in defining the heritage value of the natural resources. This is where different ways of knowing meet: the knowledge based in experience and intimate dialogue between a feature in the landscape and the native resident versus science-based knowledge that has developed through the making of hard evidence (cf. Scott, 1998). This juxtaposition serves to highlight how and why heritage in combination with the concept of environmental communication can be utilized to advance the communication of different environmental perceptions.

We return to Sweden in the chapter "'The Sea Has No Boundaries': Collaboration and Communication Between Actors in Coastal Planning on the Swedish West Coast", where Simon Larsson and Annelie Sjölander-Lindqvist study environmental communication as an intra- and inter-organizational phenomenon. They direct our attention to Sweden's West Coast, the challenges of coordinating spatial planning in a governance setting, and the difficulties involved in developing and coming to agreements about plans conceived to cope with the sustainable development of coastal and marine areas. Approaching the planning process as a continuous interaction between actors, they find that the established institutional division of roles and responsibilities, as well as current legislation, creates challenges for cooperation and producing outcomes in a collaborative governance setting. Seeing the collaboration and the dialogue as a setting where meanings both drive interaction and are themselves negotiated during such interaction, they identify a place where anthropology meets the field of environmental communication. The anthropological gaze, focusing on working beliefs rather than relying on ready-made categories to analyze social institutions, highlights how environmental communication procedures operate interactively and articulate condensed and ontologically situated meanings through administrative actions aimed at negotiated decisions.

Finally, in the chapter "Power, Conflicts, and Environmental Communication in the Struggles for Water Justice in Rural Chile: Insights from the Epistemologies of the South and the Anthropology of Power", we cross the Atlantic once again and arrive in Chile, where Cristian Alarcón takes inspiration from the anthropology of power and the epistemologies of the South in analyzing struggles for water justice and water democracy in the country. He uses this case study to argue more broadly for the relevance of a conflict- and power-oriented conceptualization of environmental communication, since this form of communication is situated in struggles around the present and the future of human–environment interaction. These struggles are implicated in a context of who is given the right to use the water: Is it the people who use it or the actor who owns the resource? As Alarcón argues, the struggles are entangled in a neoliberal conflict constituted by the negotiation of the boundaries of hegemonic political and social power and knowledge. Through the use of environmental communication, people demand a renegotiation of the structural conditions underlying and restricting the consumptive use and management of water resources. This chapter shows how environmental conflict is not only "ingrained in struggles over the meaning and materiality of social-ecological conflicts today" (p. 216), but also how environmental communication is normatively ingrained in the contingencies of epistemology, socio-ecological relations and issues of power, and discursively dispersed rights systems.

Final Note

A final note on the contributions to this book and what unites the different case studies: Environmental communication in its truest sense is made up of individuals who hold collectively shared, particular ideas, assumptions, and values, encompassing various meanings, aspirations, and intentions of those involved. Each actor, each individual, each sector, and each representative have their own set of criteria concerning what constitutes valid or valuable knowledge of aspects relating to the issue of concern in the dialogue.

In addition to being informed politically and regulatorily, environmental communication is also guided by sector- and locale-specific norms and values, as well as differently construed ideas of temporality and heritage, exposing contrasting ideas of the past, present, and future. All the case studies are also embedded in temporality, and arguments for specific actions and worldviews display time horizons that differ according to divergent understandings of science, epistemological premises, values, and value priorities. These ideas and images are concomitant with how place and landscape are tied to local history, collective memory, and knowledge, and how people, over time, by using natural resources, establish meaning-building relationships with one another and with their environments (Stoffle et al., 2013). The dimension of time is showing itself to be a potentially critical variable in understanding what makes up environmental communication.

The issue of epistemology is also a pertinent theme in the book. In the coming chapters, we will see divides between science on the one hand and Local, Traditional, and Indigenous knowledge on the other, each way of knowing grounded in different epistemological and ontological assumptions. The assumed objectivity of science tends to give it a powerful voice to speak for the environment and how it should be managed

(Sjölander-Lindqvist et al., 2020b), thereby normalizing particular policies for environmental management and authorizing certain experts to act in management (Goldman et al., 2011). The result tends to be the establishment of a knowledge hierarchy where, for example, experience-based knowledge is considered subordinate and local livelihood-based discourses are dismissed, while scientific models and experts' understanding are seen as providing superior knowledge for handling pressing issues (Agrawal, 2005; Sjölander-Lindqvist, 2008) as well as overall environmental interpretation despite conventions and agreements to include different ways of knowing (Sjölander-Lindqvist et al., 2020b). Knowledge is henceforth both repressive and productive, and environmental communication is a site of power where truths are made, circulated, and remade. It has elsewhere been discussed how policies have "social lives of their own" (Sjölander-Lindqvist, 2015; cf. Appadurai, 1986), and the same can be argued in the case of environmental communication, as it is a process through which knowledge, interests, and values are constantly upheld, demarcated, and negotiated.

This emphasizes the plurality of the concept of environmental communication itself. In 2010, Chris Shore, an anthropologist dedicated to the study of policies and policy work, defined the task of anthropology as examining practices "in work" and focusing on "the conditions that create and sustain them and the kinds of relations and subjects they produce" (p. 213; cf. Shore et al., 2011). I would like to borrow this thought from Shore and open the floor for further anthropological engagement with environmental communication from the perspective that anthropology can, and should, approach environmental communication as a cultural process that occurs in different contexts.

Any critical analysis of communication at the nexus of the human and the environmental, and the practices undertaken within this relational node, involves capturing and representing the meanings of particular situations and clarifying their conditions and unique circumstances. This requires sensitivity to the tangible and associative values of those concerned and involved, and to the circulating discourses, multiple contestations, and regimes of power enacted and confirmed within the field of environmental communication.

This requires the ability and the patience to become familiar with the realities of a bounded cultural setting in order to describe it and proceed with analysis. And that involves speaking with people about their way of perceiving events and, as the various cases in this book demonstrate, interrogating their ideas and understandings of the material and the immaterial, the exceptional and unusual occurrences, as well as the ordinary occurrences of everyday life. That is, life and environmental communication as lived, interpreted, and given meaning. Such a perspective lends importance to how ethnographic methods, with their holistic scope, can contribute to the vital acknowledgment of embedded, locally specific perceptions in the narratives shared during fieldwork. We are thus deeply indebted to the field locations, informants, and all the other participants that make anthropological inquiry possible.

Acknowledgments I am grateful to Anna Bohlin, Maris Boyd Gillette, Simon Larsson, and Richard Stoffle for their very useful comments and suggestions on earlier versions of this introduction. I would also like to thank all the authors in this volume for their inspiring ideas and comments on anthropology and environmental communication. Last, but definitely not least, I would like to take the opportunity to express my and fellow authors' gratitude to all those who have shared their time and thoughts with us during our fieldwork.

References

Abram, D. (1996). *The spell of the sensuous: Perception and language in a more-than-human world.* Pantheon Books.

Agrawal, A. (2005). *Environmentality: Technologies of government and the making of subjects.* Duke University Press.

Alarcón, C. (2015). *Forests at the limits.* Doctoral thesis, Swedish University of Agricultural Sciences. Retrieved from https://pub.epsilon.slu.se/11926/1/alarcon_ferrari_c_150223.pdf

Anderson, E. N. (2014). *Caring for place: Ecology, ideology, and emotion in traditional landscape management.* Left Coast Press.

Appadurai, A. (Ed.). (1986). *The social life of things: Commodities in cultural perspective.* The Press Syndicate of the University of Cambridge.

Armitage, D., Charles, A., & Berkes, F. (2017). *Governing the Coastal Commons: Communities, resilience and transformation.* Routledge.

Armstrong, A. (2006). Ethical issues in water use and sustainability. *Area, 38*(1), 9–15. https://doi.org/10.1111/j.1475-4762.2006.00657.x

Bassiouney, R. (Ed.). (2018). *Identity and dialect performance: A study of communities and dialects.* Routledge.

Bate, S. P. (1997). Whatever happened to organizational ethnography? A review of the field of organizational ethnography and anthropological studies. *Human Relations, 50*(9), 1147–1175.

Bateson, G. (1972). *Steps to an ecology of mind: Collected essays in anthropology, psychiatry, evolution, and epistemology*. Jason Aronson.
Blythe, J., Silver, J., Evans, L., Armitage, D., Bennett, N. J., Moore, M.-L., Morrison, T. H., & Brown, K. (2018). The dark side of transformation: Latent risks in contemporary sustainability discourse. *Antipode, 50*(5), 1206–1223.
Boholm, Å. (2015). *Anthropology and risk*. Earthscan.
Brady, M. J. (2011). Mediating indigenous voice in the museum: Narratives of place, land, and environment in new exhibition practice. *Environmental Communication, 5*(2), 202–220. https://doi.org/10.1080/17524032.2011.562649
Brevini, B. (2016). The value of environmental communication research. *The International Communication Gazette, 78*(7), 684–687. https://doi.org/10.1177/1748048516655728
Brondizio, E. S., Ostrom, E., & Young, O. R. (2009). Connectivity and the governance of multilevel social-ecological systems: The role of social capital. *Annual Review of Environment and Resources, 34*(1), 253–278.
Chaffe, B. C., Garmestani, A. S., Gunderson, L. H., Benson, M. H., Angeler, D. C., Arnold, C. A., Cosens, B., Craig, R. K., Ruhl, J. B., & Allen, C. R. (2016). Transformative environmental governance. *Annual Review of Environment and Resources, 41*(1), 399–423.
Cox, R. (2013). *Environmental communication and the public sphere*. SAGE.
Cox, R., & Depoe, S. (2015). Emergence and growth of the "field" of environmental communication. In A. Hansen & R. Cox (Eds.), *The Routledge handbook of environment and communication* (pp. 13–25). Routledge.
Dávid-Barrett, T., & Dunbar, R. (2016). Language as a coordination tool evolves slowly. *Royal Society Open Science, 3*, 160259.
Díaz, S., Settele, J., Brondízio, E. S., Ngo, H. T., Agard, J., Arneth, A., Balvanera, P., Brauman, K. A., Butchart, S. H. M., Chan, K. M. A., Garibaldi, L. A., Ichii, K., Liu, J., Subramanian, S. M., Midgley, G. F., Miloslavich, P., Molnár, Z., Obura, D., Pfaff, A., Polasky, S., Purvis, A., Razzaque, J., Reyers, B., Chowdhury, R. R., Shin, Y.-S., Visseren-Hamakers, I., Willis, K. J., & Zayas, C. N. (2019). Pervasive human-driven decline of life on Earth points to the need for transformative change. *Science, 366*(6471), eaax3100. https://doi.org/10.1126/science.aax3100
Escobar, A. (2001). Culture sits in places: Reflections on globalism and subaltern strategies of localization. *Political Geography, 20*(2), 139–174. https://doi.org/10.1016/S0962-6298(00)00064-0
Evans Comfort, S., & Park, Y. P. (2018). On the field of environmental communication: A systematic review of the peer-reviewed literature. *Environmental Communication, 12*(7), 862–875. https://doi.org/10.1080/17524032.2018.1514315

Fitzgerald, T. K. (1993). *Metaphors of identity: A culture-communication dialogue.* State University of New York Press.
Fjællingsdal, K. S., & Klöckner, C. A. (2020). Green across the board: Board games as tools for dialogue and simplified environmental communication. *Simulation & Gaming, 51*(5), 632–652. https://doi.org/10.1177/1046878120925133
Geertz, C. (1973). *The interpretation of cultures.* Basic Books.
Geertz, C. (1983). *Local knowledge: Further essays in interpretative anthropology.* Basic Books.
Goldman, M. J., Nadasdy, P., & Turner, M. D. (Eds.). (2011). *Knowing nature: Conversations at the intersection of political ecology and science studies.* The University of Chicago Press.
Hammersley, M., & Atkinson, P. (1995). *Ethnography: Principles in action* (2nd ed.). Routledge.
Hansen, A., & Cox, R. (2015). Introduction: Environment and communication. In A. Hansen & R. Cox (Eds.), *The Routledge handbook of environment and communication* (pp. 1–10). Routledge.
Harvey, P., & Know, H. (2015). *Roads: An anthropology of infrastructure and expertise.* Cornell University Press.
Heatherington, T. (2010). *Wild Sardinia: Indigeneity & the global dreamtimes of environmentalism.* University of Washington Press.
Hylland Eriksen, T. (1991). The challenges of anthropology. *International Journal of Pluralism and Economics Education, 1*(3), 194–202.
Ingold, T. (2000). *The perception of the environment: Essays in livelihood, dwelling and skill.* Routledge.
Ingold, T. (2008). Anthropology is not ethnography. *Proceedings of the British Academy, 154*, 69–92.
Ingold, T. (2010). Anthropology comes to life. *General Anthropology, 17*(1), 1–4.
James, A., Hockey, J., & Dawson, A. (1997). *After writing culture: Epistemology and praxis in contemporary anthropology.* Routledge.
James, A., & Marcus, A. (2006). The human role in changing fluvial systems: Retrospect, inventory and prospect. *Geomorphology, 79*(3), 152–171.
Joosse, S., Powell, S., Bergeå, H., Böhm, S., Calderón, C., Caselunghe, E., Fischer, A., Grubbström, A., Hallgren, L., Holmgren, S., Löf, A., Nordström Källström, H., Raitio, K., Senecah, S., Söderlund Kanarp, C., von Essen, E., Westberg, L., & Westin, M. (2020). Critical, engaged and change-oriented scholarship in environmental communication: Six methodological dilemmas to think with. *Environmental Communication, 14*(6), 758–771. https://doi.org/10.1080/17524032.2020.1725588
Jordan, P. (2003). *Material culture and sacred landscape: The anthropology of the Siberian Khanty.* Altamira Press.

Katz, C. (1992). All the world is staged: Intellectuals and the projects of ethnography. *Environment and Planning D: Society and Space, 19*, 495–510.

Kellert, S. R., Mehta, J. N., Ebbin, S. A., & Lichtenfeld, L. L. (2000). Community natural resource management: Promise, rhetoric, and reality. *Society & Natural Resources, 13*(8), 705–715. https://doi.org/10.1080/089419200750035575

Kopnina, H., & Shoreman-Ouimet, E. (Eds.). (2013). *Environmental anthropology: Future directions*. Routledge.

Kottak, C. P. (2010). *Anthropology: Appreciating human diversity* (14th ed.). McGraw-Hill.

Krauss, W., & von Storch, H. (2012). Post-normal practices between regional climate services and local knowledge. *Nature and Culture, 7*(2), 213–230. https://doi.org/10.3167/nc.2012.070206

Lakoff, G., & Johnson, M. (2003[1980]). *Metaphors we live by*. Chicago: University Chicago Press.

Lam, D., Hinz, E., Lang, D., Tengö, M., von Wehrden, H., & Martín-López, B. (2020). Indigenous and local knowledge in sustainability transformations research: A literature review. *Ecology and Society, 25*(1), 3. https://doi.org/10.5751/ES-11305-250103

Leach, E. (1976). *Culture and communication: The logic by which symbols are connected*. Cambridge University Press.

LeCompte, M. D., & Schensul, J. J. (2013). *Analysis and interpretation of ethnographic data: A mixed methods approach*. AltaMira Press.

MacLeod, M., & Macdonald, I. (2018). Learning from the locals: How can co-design support malaria education in a post-colonial environment? *Design for Health, 2*(1), 163–185. https://doi.org/10.1080/24735132.2018.1451676

McAfee, D., Doubleday, Z. A., Geiger, N., & Connell, S. D. (2019). Everyone loves a success story: Optimism inspires conservation engagement. *BioScience, 69*(4), 274–281. https://doi.org/10.1093/biosci/biz019

Mesoudi, A., Whiten, A., & Dunbar, R. (2006). A bias for social information in human cultural transmission. *British Journal of Psychology, 2006*, 405–423.

Morin, O. (2016). *How traditions live and die*. Oxford University Press.

Morrison, G. (2018). *Art and the environment: Museums adjust to a new climate*. Last retrieved February 14, 2021, from https://www.sothebys.com/en/articles/art-and-the-environment-museums-adjust-to-a-new-climate

Moser, S. C. (2016). Reflections on climate change communication research and practice in the second decade of the 21st century: What more is there to say? *WIREs Climate Change, 7*(3), 345–369. https://doi.org/10.1002/wcc.403

Nguyen, M. T. N. (2016). Trading in broken things: Gendered performances and spatial practices in a northern Vietnamese rural-urban waste economy. *American Ethnologist, 43*(1), 116–129. https://doi.org/10.1111/amet.12267

Paerregaard, K. (2020). Communicating the inevitable: Climate awareness, climate discord, and climate research in Peru's highland communities.

Environmental Communication, 14(1), 112–125. https://doi.org/10.1080/175240.32.2019.1626754
Pascual, U., Phelps, J., Garmendia, E., Brown, K., Corbera, E., Martin, A., Gomez-Baggethun, E., & Muradian, R. (2014). Social equity matters in payments for ecosystem services. *BioScience, 64*(11), 1027–1036.
Pink, S. (2012). *Situating Everyday Life: Practices and Places.* London: Sage.
Primmer, E., & Kyllönen, S. (2006). Goals for public participation implied by sustainable development, and the preparatory process of the Finnish National Forest Programme. *Forest Policy Economics, 8*, 838–853. https://doi.org/10.1016/j.forpol.2005.01.002
Risvoll, C., & Kaarhus, R. (2020). Struggling with 'clear zoning': Dilemmas of carnivore-pastoral coexistence in Nordland, northern Norway. In A. Breilid & R. Krøvel (Eds.), *Indigenous knowledges and the sustainable development agenda* (pp. 185–206). Routledge.
Rival, L. (Ed.). (2001). *The social life of trees: Anthropological perspectives on tree symbolism* (2nd ed.). Berg.
Romero, S. (2015). *Language and ethnicity among the K'ichee' Maya.* The University of Utah Press.
Sandström, C., Sjölander-Lindqvist, A., Pellikka, J., Hiedanpää, J., Krange, O., & Skogen, K. (2018). Between politics and management: Governing large carnivores in Fennoscandia. In T. Hovardas (Ed.), *Large carnivore conservation and management: Human dimensions* (pp. 269–290). Routledge.
Schensul, J. J., & LeCompte, M. D. (2016). *Ethnography in action: A mixed methods approach.* Rowman & Littlefield.
Scott, J. C. (1998). *Seeing like a state.* Yale University Press.
Semanza, J. C., Hall, D. E., Wilson, D. J., Bontempo, B. D., Sailor, D. J., & George, L. A. (2008). Public perception of climate change: Voluntary mitigation and barriers to behavior change. *American Journal of Preventive Medicine, 35*(5), 479–487. https://doi.org/10.1016/j.amepre.2008.08.020
Senecah, S. (2004). *The environmental communication yearbook.* Lawrence Erlbaum Associated Publishers.
Shore, C. (2010). Locating the work of policy. In H. Colebatch, R. Hoppe, & M. Noordegraaf (Eds.), *Working for policy* (pp. 211–226). Amsterdam University Press.
Shore, C., Wright, S., & Però, D. (Eds.). (2011). *Policy worlds: Anthropology and analysis of contemporary power.* Berghahn.
Sjölander-Lindqvist, A. (2004). Visualizing lace and belonging: Landscape redefined in a Swedish farming community. In Å. Boholm & R. E. Lofstedt (Eds.), *Facility siting: Risk, power and identity in land-use planning.* Earthscan.
Sjölander-Lindqvist, A. (2008). Identity, science and politics indivisible: The Swedish wolf controversy deconstructed. *Journal of Environmental Policy and Planning, 10*(1), 71–94.

Sjölander-Lindqvist, A. (2009). Social-natural landscape reorganised: Swedish forest-edge farmers and wolf recovery. *Conservation and Society, 7*(2), 130–140.

Sjölander-Lindqvist, A. (2015). Balancing differentiated interests and conceptualizations in environmental management. *Journal of Organizational Ethnography, 4*(3), 306–323.

Sjölander-Lindqvist, A., Skoglund, W., & Laven, D. (2020a). Craft beer – building social terroir through connecting people, place and business. *Journal of Place Management and Development, 13*(2), 149–162.

Sjölander-Lindqvist, A., Risvoll, C., Kaarhus, R., Lundberg, A. K., & Sandström, C. (2020b). Knowledge claims and struggles in decentralized large carnivore governance: Insights from Norway and Sweden. *Frontiers in Ecology and Evolution, 8*, 120. https://doi.org/10.3389/fevo.2020.00120

Sobo, E. J., & de Munck, V. C. (1998). The forest of methods. In V. C. de Munck & E. J. Sobo (Eds.), *Using methods in the field: A practical introduction and casebook* (pp. 13–37). Altamira Press.

Spence, A., Poortinga, W., & Pidgeon, N. (2012). The psychological distance of climate change. *Risk Analysis, 32*(6), 957–972. https://doi.org/10.1111/j.1539-6924.2011.01695.x

Stewart, O., Lewis, H., & Anderson, K. (2002). *Forgotten fires: Native Americans and the transient wilderness*. University of Oklahoma Press.

Stoffle, R. W., Stoffle, B. R., & Sjölander-Lindqvist, A. (2013). Contested time horizons. In A. Bond, A. Morrison-Saunders, & R. Howitt (Eds.), *Sustainability assessment: Pluralism, practice and progress* (pp. 51–67). Routledge.

Strang, V. (2004). *The meaning of water*. Berg.

Sygna, L., O'Brien, K., & Wolf, J. (2013). *A changing environment for human security: Transformative approaches to research, policy, and action*. Earthscan.

Thomas, W. L., Sauer, C. O., Bates, M., & Mumford, L. (1956). *Man's role in changing the face of the earth*. University of Chicago Press.

Titsworth, S., Hosek, A., Pearson, J., & Nelson, P. (2021). *Human communication* (7th ed.). McGraw Hill.

Toda, M. (1976). The decision process: A perspective. *International Journal of General Systems, 3*(2), 79–88.

Turner, V. (1977). Process, system, and symbol: A new anthropological synthesis. *Discoveries and Interpretations: Studies in Contemporary Scholarship, 106*(3), 61–80.

UN. (2015). *Transforming our world: The 2030 agenda for sustainable development*. United Nations.

Underhill, J. W. (2011). *Creating worldviews: Metaphor, ideology and language*. Edinburgh University Press.

Valadez, J. M. (2018). *Deliberative democracy, political legitimacy, and self-determination in multicultural societies*. Routledge.

Walker, J. M. M., Godley, B. J., & Nuno, A. (2019). Media framing of the Cayman Turtle Farm: Implications for conservation conflicts. *Journal for Nature Conservation, 48*, 61–70. https://doi.org/10.1016/j.jnc.2019.01.001

Wassef, N. (2001). On selective consumerism: Egyptian women and ethnographic representations. *Feminist Review, 69*, 111–123. https://doi.org/10.1080/014177800110070148

Worth, S., & Adair, J. (1975). *Through Navajo eyes: An exploration in film communication and anthropology*. Indiana University Press.

Zhouri, A. (2018). Megaprojects, epistemological violence and environmental conflicts in Brazil. *Perfiles Económicos, 5*, 7–33.

Zikargae, M. H. (2018). Analysis of environmental communication and its implication for sustainable development in Ethiopia. *Science of the Total Environment, 634*, 1593–1600. https://doi.org/10.1016/j.scitotenv.2018.04.050

Dancing with Lava: Indigenous Interactions with an Active Volcano in Arizona

Kathleen Van Vlack

Introduction

The Little Springs Lava Flow is located in northern Arizona near the north rim of the Grand Canyon in the American Southwest (Fig. 1). This lava flow is part of a larger volcanic landscape known as the Uinkaret Lava Field. The human connections and usage of this unique landscape are at the center of a heritage environmental communication debate. This debate is centered on creating meaning regarding places, objects, and landscapes, and it shapes how people see and value heritage. It also can bring to the forefront the conflicting views on heritage places and landscapes (Hanson & Cox, 2015). This chapter highlights the debate between the Southern Paiute tribes and the archeologists who conducted field surveys in the area over how and when the Little Springs Lava Flow was used. This debate affects which heritage interpretation contributes to heritage management decisions and cultural displays and signage.

K. Van Vlack (✉)
Living Heritage Research Council, Cortez, CO, USA

A. Sjölander-Lindqvist et al. (eds.), *Anthropological Perspectives on Environmental Communication*, Palgrave Studies in Anthropology of Sustainability, https://doi.org/10.1007/978-3-030-78040-1_2

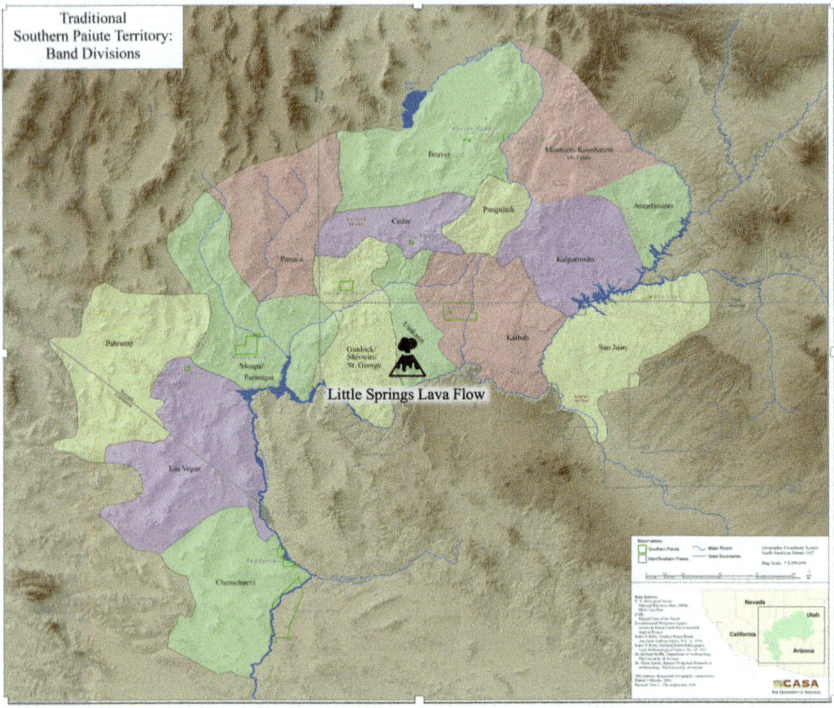

Fig. 1 Little Springs Lava Flow located within traditional Southern Paiute territory. (This image used with permission of Richard Stoffle)

This chapter brings forth a particular environmental communication issue that has been noted by Joosse et al. (2020). They note that representing marginalized voices is key for change-oriented and engaged environmental communication scholarship and that this type of research can offer different perspectives on contested environmental issues (Joosse et al., 2020; Milstein & Kroløkke, 2012). Voices are excluded from discussions of heritage and environmental conservation due to gender, social class, race, and lack of ability to verbally speak, and this is the result of power relations that have become deeply entrenched in society. Joosse et al. (2020) pointed out that environmental communication scholars in recent years have highlighted this issue and note that researchers aim to give voice to those who have been historically excluded but there is a risk of silencing perspectives and experiences that do not fit the aims of the research process and goals. When environmental communication scholars challenge and counter flawed representations, they risk dealing with the long-term effects of discursive colonialism, which is the reproduction of interests of powerful state actors through various forms of knowledge and scholarship (Mohanty, 1984).

In this era of decolonization and self-determination, Southern Paiute people, like indigenous peoples around the world, are working toward reclaiming the telling of their histories and their cultures. Indigenous peoples are positioning themselves to put forth certain parts of their cultures that have been ignored or misinterpreted by heritage researchers and land managers to combat stereotypes and rhetoric that disconnect them from their traditional lands. In some places, indigenous peoples have been successful in this process, such as the Maori of New Zealand (Smith, 2012). However, in other places like the United States, it is an ongoing struggle. In the Great Basin and Colorado Plateau region of the United States, Southern Paiute people have long expressed a need to share various elements of their culture to provide an alternative perspective to counter the stories presented by archeologists and some cultural anthropologists.

Methodology

This analysis is based on multiple Native American cultural resource studies in the Grand Canyon–Arizona Strip region. The data used in this chapter was collected during two federally funded Native American ethnographic studies. The first project was funded by the Bureau of Land Management (BLM), Arizona Strip Field Office in 2003. This study focused on understanding Southern Paiute cultural landscapes, places, and cultural resources throughout the Arizona Strip (Stoffle et al., 2005). During this study, tribal representatives from the Kaibab Band of Paiute Indians along with ethnographers from the University of Arizona (UofA) first visited the Little Springs area. During this time a recommendation was made for a second study focused solely on Southern Paiute connections and interactions with the Little Springs Lava Flow.

Based on those tribal recommendations, the second follow-up study was funded in 2012 by Grand Canyon–Parashant National Monument, which is jointly managed by the BLM and the National Park Service. This follow-up study allowed for the UofA research team to bring many of the same tribal representatives who took part in the original field visits in 2003 back to the Little Springs area for a more in-depth examination of the lava flow and surrounding cultural resources (Van Vlack et al., 2013).

During both studies, the UofA team conducted interviews with tribal representatives using survey instruments that had been developed and refined over the past 25 years with the assistance of official tribal representatives, and these forms have been approved by participating tribal governments. These instruments have been administered to Indian people during at least 20 different projects since 1997. This equates to over one thousand interviews with Indian people.

During the 2012 Little Springs study, the UofA team used two survey instruments to interview tribal representatives—the Site Evaluation Form and the Cultural Landscape form. Each one of the forms provides Indian people an opportunity to discuss the cultural importance of a place and how it is connected to other areas within their traditional territory and the surrounding local landscape. In total 58 interviews were conducted in the Little Springs area during the two studies.

Previous Research

There are only a few published studies that attempt to understand the human connections and uses of this unique landscape. Most have been archeological studies (Elson & Ort, 2006; Hintzman, 2012; Ort et al., 2008a, 2008b). These studies have used the available archeological data and information on the volcanic eruption to interpret the American Indian lifestyles and behavioral responses that occurred before, during, and after the Little Springs event. Two ethnographic studies have been published involving the Kaibab Band of Paiute Indians; these reports documented the Southern Paiute interpretation of the lava flow and how it is important to the larger Grand Canyon landscape (Stoffle et al., 2005; Van Vlack et al., 2013).

Most of the previous accounts agree that Indian people used the Uinkaret Volcanic Field area for long periods before the Little Springs volcanic event. Many scholars also agree that the Little Springs event was special because Indian people came to interact with a hornito in the active lava flow and made sherd rocks, and that Indian people made up to five miles of trails on the two lava flows and hundreds of round low wall structures. A rift occurs, however, when archeologists and Southern Paiutes answer the questions of who the Indian people who used this area were and for what purpose the Indian people used the lava flow.

The Southern Paiute interpretation of the significance of the Little Springs Lava Flow and the Uinkaret Volcanic Field varies greatly from the published archeological interpretations. While Southern Paiutes and archeologists are in agreement that Indian people constructed the trails and associated structures, the archeologists have maintained that this area was not a ceremonial area as Southern Paiute representatives have stated, but instead it was a place Indian people fled from during the time of the eruption (Elson & Ort, 2006; Hintzman, 2012; Ort et al., 2008a, 2008b). Archeologists also have argued that Indian people returned to the area post-eruption to live on top of the lava flow for protection. Structures and trails were built for defensive purposes to protect against an unknown enemy.

Southern Paiute Epistemology

To understand the Southern Paiute perspective on how places and resources are ascribed meaning, it is essential to examine Southern Paiute epistemology. As explained by Liljeblad (1986, pp. 643–644), to the Southern Paiutes, "Puha is everywhere and is a source of individual competence, mental and physical ability, health, and success." This concept is common amongst many different tribes living throughout the western United States such as the Ute, Western Shoshone, Owens Valley Paiutes, Northern Paiute, Mojave, Hualapai, and Havasupai. *Puha* is a fundamental principle of their epistemologies as well.

Southern Paiute people's belief in a living universe and the notion that everything has *Puha* shapes how they engage with the landscape and how they approach the act of pilgrimage. This epistemological foundation of Southern Paiute culture is similar to what anthropologist Roy Rappaport (1999, pp. 263–71 and p. 446) calls an *ultimate sacred postulate* and what social philosopher Alvin Goldman (1999) calls a philosophical *primitive*.

In Southern Paiute culture, pilgrimage has always focused on the acquisition of *Puha*, the spiritual transformation of *Puha'gants* (Southern Paiute medicine men) and building relationships and communities. *Puha* is derived from Creation and permeates the universe, which resembles a spiderweb. Sometimes it is like a thin scattering; at other times, it occurs where there are clusters of life in high concentrations. *Puha* exists throughout the universe but varies in intensity from person to person, place to place, element to element, and object to object, similar to how strength differs among humans (Van Vlack, 2012a, 2012b).

Southern Paiute people maintain that they are the people who were created in and for this cultural landscape. According to Southern Paiute oral history, when the world was formed, the Creator established a set of mutual obligations and responsibilities between Southern Paiute people and this living natural world. As a result, places like volcanoes (Little Springs) need to be interacted with, maintained, and respected.

The Little Springs Lava Flow was an important ceremonial area on the North Rim of the Grand Canyon during the eruption, and Southern Paiutes later returned to construct miles of cinder trails and structures to use in post-eruption ceremonies. It is also estimated that there are at least eight kilometers of trails located on the two lava lobes. Indian people also made hundreds of circular structures along these trails. Evidence suggests that this area became a major destination for Paiutes seeking healing and spiritual enrichment.

Use of the Uinkaret Volcanic Field Before the Little Springs Event (17,000 B.P. to A.D. 1075)

Most scholars agree that American Indian people were living in the Mount Trumbull and Uinkaret Volcanic Field before the Little Springs volcanic event. Paleo-Indian time-period spears, dart points, and arrow points have been found throughout the Arizona Strip, documenting at least 12,000 years of continuous occupation and use by Indian people. For most of this long period, there was no agriculture north of the Colorado River, so the most important American Indian subsistence activities were hunting, gathering of wild plants, and the management of semi-domestic cultigens.

In the thousands of years before A.D. 500, there is evidence (Rhode & Louderback, 2007) that Indian people used hundreds of food, medicine, and weaving plants, including amaranth, *Chenopodium*, *Oryzopsis hymenoides*, *Stanleya pinnat*, and *Prosopis glandulosa* for seed production; *Ephedra nevadensis* for medicine; and *Harpagophytum procumbens* for weaving. After using these plants in the same ecosystems for thousands of years, Indian people learned how to encourage and modify many of these plants. Horticulture, defined here as conscious and systematic plant manipulation, thus developed over time due to learning about the physical and biological characteristics of the land and plants. This is often referred to as traditional ecological knowledge (TEK), and experimentation with various land-use strategies helped make both land and plants better able to serve Indian people.

There were, however, no major cultigens like corn, beans, or squash north of the Colorado River before this time (A.D. 500) and no evidence of large-scale irrigated agriculture. By A.D. 700, agriculture had diffused throughout the region. Agricultural production in this region rapidly rose in complexity and spatially expanded, corresponding with a climatic shift involving temperature warming and increased rainfall. Pottery production and its elaboration seemed to be directly tied with the more stable community life that is commonly associated with irrigation farming. This emergent lifestyle resulted in a steadily increasing population, a more sedentary lifestyle, the development of social hierarchy, the elaboration of pottery, and probably other elements of culture that do not appear in the archeological record.

The peak of this irrigated-agriculture-based way of life occurred between A.D. 1000 and A.D. 1275. Soon, however, the Little Ice Age, which is currently dated at A.D. 1275 to 1850 and now appears to have been caused by volcanic eruptions elsewhere, marked another major climatic shift. Regional lifeways in the Arizona Strip would rapidly change again due to the onset of the Little Ice Age: by 1300, the climate in the region had drastically changed and the medium-to-small side creeks, which were needed for much of the irrigated agriculture, had dried up, and only the mainstream portion of rivers remained. The UofA team estimates that by 1300 up to 79% of the preexisting irrigated agriculture in the Kaibab Paiute district was no longer possible because of greater aridity and major changes to weather patterns. As an adaptation to the new climatic conditions, Indian people shifted to a more reliable and mobile lifestyle, which was dependent on the use of natural resources and semi-domestic cultigens, although irrigated agriculture continued to be conducted in the mainstream rivers. Pottery was less useful in this less sedentary lifestyle, so basketry was greatly elaborated. Their focus shifted from pretty pots to beautiful baskets (Van Vlack et al., 2013).

It is unlikely that there were the numbers and kinds of religious specialists in the Mount Trumbull and Uinkaret Volcanic area needed for a reasoned response to the volcanic event. If the latter occurred, then it is possible that religious specialists who were focused on volcanic ceremonies could have been in place near Mount Trumbull at the time of the volcanic event. A second question is "How fast did the eruption event occur?" If the full event occurred over a short period (say a few days or a week) then the local people in place at the time would have had little time to make a ceremonial response to the event. If the full event lasted over a long period (say a few months) then time would have been available for runners to be sent to distant Southern Paiute and other Indian residential areas to inform them of the impending event, and they consequently would have had the time to come to the Uinkaret Volcanic Field to discuss appropriate ceremonial responses and carry out these responses during the event itself and after it.

Based on ethnographic current research on the aboriginal water uses of the Kaibab Paiute people, previous research on the aboriginal water uses of the Shivwits/Santa Clara Paiutes (Stoffle et al., 1994), and research on Zion National Park (Stoffle et al., 1997), it is clear that the Uinkaret Volcanic Field was an ecologically marginal area for farming. Instead, for all three Paiute districts, the most productive irrigated agricultural fields were concentrated at much lower elevations along permanent water sources. Each district also had one or more upland or hinterland areas where various activities occurred which could include ceremonies. The irrigated farming areas and primary residential areas for the people of the Uinkaret/Uyu'naits district were along the lower reaches of the Virgin River and its immediate tributaries at an elevation of about 1158 meters. Whatever activities occurred in the Uinkaret/Uyu'naits hinterland, they did so at an elevation of at least 1981 meters.

Given the marginal farming conditions available on the Uinkaret Volcanic Field, it is more likely, based on Southern Paiute cultural logic, that the Uinkaret/Uyu'naits volcanic landscape was primarily used as an important ceremonial landscape rather than a place where many people lived year-round. Those people that did live year-round or most of the year in this landscape were probably Paiute religious leaders and their families who were here to support the conduct of ceremonies related to both this volcanic landscape and powerful places located to the south along the Colorado River. It thus is likely that this hinterland was primarily used for ceremony by a wide variety of religious leaders from many Paiute and other ethnic groups.

Farming in the Uinkaret/Uyu'naits volcanic landscape certainly did occur on a small scale; small number of people who lived in this powerful isolated area for ceremonial purposes practiced some form of farming. The residential Uinkaret/Uyu'naits religious leaders and their families probably farmed here to support themselves and visiting groups seeking places for ceremony rather than the farming here serving as the primary food source for the majority of the Uinkaret/Uyu'naits people.

Little Springs Volcanic Event and the Southern Paiute Response

The Southern Paiute interpretation is that when the Little Springs eruption began building up, Paiute Puha'gants began monitoring the seismic activity in the area. The presence of the sherd rocks (pottery pieces fused into volcanic rock) argues for a well-planned and culturally reasoned response to the volcanic event. Such reasoning about the possibility of an eruption event would have to have been conducted over considerable time by religious leaders. The sufficient geological warnings occurred over a lengthy period of an impending eruption, which allowed for local Paiute leaders to send out runners to announce the potential of a coming volcanic event and for Paiute volcanic specialists from great distances to travel to the Mount Trumbull area. After religious leaders were gathered, they would have had time to work together on a reasoned response to a potential volcanic event. This is in contrast with the views of archeologists and other researchers. According to Northern Arizona University volcanologist Michael Ort, the entire Little Springs volcanic event could have happened over a span of as little as a few days to a week (Ort et al., 2008b). If true, this would not have allowed the Indian people to prepare a response in anticipation of the event.

A key factor in determining how Southern Paiute people would have responded to a volcanic event and volcanoes is understanding Paiute environmental knowledge surrounding volcanism. It is known that Southern Paiute people had dozens of kinds of medical specialists and shamans (Kelly, 1939), some of whom used volcanic materials such as obsidian in their religious practices. Omer Stewart (1942, p. 305) documented in the 1930s that Southern Paiute mothers would feed their babies the breast milk that had been placed on a volcanic rock to give the milk and thus the child strength. It is also known that Southern Paiute people, like most Indian ethnic groups in the western US, went to volcanoes to seek knowledge, heal the sick, and to conduct regional and world balancing ceremonies (Stoffle et al., 2015).

Contemporary Southern Paiute people believe that Paiute people possessed sufficient traditional ecological knowledge (TEK) to predict the Little Springs volcanic event and to plan a specific set of behavioral responses once it occurred. The most likely explanation for this is that the Little Springs event happened soon after the massive and lengthy Sunset Crater volcanic eruption. UofA researchers have worked with cultural representatives from eight American Indian tribes regarding their interaction with and interpretation of the Sunset Crater event (Toupal & Stoffle, 2004). While the relative dates for the two events have not been established, it is known that the Sunset Crater event lasted much longer (up to a generation) than the Little Springs event. It is also known from Elson and Ort's research that during the Sunset Crater event, Indian people interacted with the active lava flows and made corn rocks (lava rocks with impressions of corn cobs in them) along edges of the hornito walls (Elson et al., 2002). These rocks then were incorporated into nearby structures, which appear to have been built for conducting ceremonies. Sunset Crater was where these religious innovations first occurred. There at Sunset Crater, and no time before to our knowledge, religious leaders developed this unique manner of interacting with magma flows to produce corn rocks that subsequently were used in making nearby ceremonial structures. This volcanic TEK was the intellectual foundation for the Indian cultural responses to the Little Springs volcanic event.

Parallels can be drawn between the Little Springs volcanic event and the ceremonial activities that occurred at the earlier volcanic episode at Sunset Crater. As with the Little Springs eruption, Indian people came to Sunset Crater to interact with, observe, and study the volcanic event. This type of engagement allowed for religious specialists and leaders from numerous ethnic groups, including Southern Paiute, to gain valuable volcano traditional ecological knowledge (TEK). It can be argued that since Sunset Crater occurred before the Little Springs eruption, Indian people used the knowledge gained at Sunset Crater to safely engage the Little Springs Lava Flow physically and spiritually.

In order to learn about the flowing lava, these religious specialists approached it in the same manner as that described previously for the Little Springs Lava Flow. Instead of using pottery to create new ceremonial items, at Sunset Crater, they used corncobs. The corn was placed on the hornito rims and corn rocks were created when magma that was forced upward landed on the hornito rims and covered the corncobs. These newly created corn rocks were then taken to specially constructed buildings at Wupatki and incorporated into the structure walls for ceremonial purposes (Toupal & Stoffle, 2004).

At Little Springs, Southern Paiute religious specialists took the newly formed sherd rocks (Fig. 2) to an area known as the Lightning Site (Fig. 3) to use in ceremonial activities. The Puha'gants incorporated them into the structure found at this site. The walls where the rocks were found face the Northern Lobe. The placement and location of the sherd rocks likely contributed to the site's overall ceremonial function.

The Lightning Site sits on top of a roughly 3.5-million-year-old basalt lava flow. This lava originated from nearby Mount Trumbull, which is located less than two kilometers to the north. The volcanic activity in this area connects places, people, and objects, and the underground magma flows carry Puha and spiritual beings to and from areas in and around the Little Springs Lava Flow. Present at this site are approximately seven to ten room blocks with upright slabs. The room blocks form a C-shape, with a center plaza that faces to the east. The room blocks have two prominent gaps between them in a spot where one would have a view to the west of the Little Springs Lava Flow. Two grinding stones were also found near the room blocks, which may have been used to prepare food items or medicines to be used in ceremonies that involved the sherd rocks.

When Elson and Ort (2006) surveyed this area, they noted that six sherd rocks were found at this site when they surveyed it in the early 2000s. The sherd rocks contain distinct pieces of pottery which were

Fig. 2 Sherd rocks. (Photo: Kathleen Van Vlack)

Fig. 3 The Lightning Site. (Photo: Kathleen Van Vlack)

identified to be Hurricane Gray and Hurricane Black-on-Gray. When visiting this site in 2012, the UofA team and Southern Paiute representatives also found other types of pottery at this site, such as Hurricane Black-on-Gray Corrugated wares. These pottery styles have been dated between A.D. 1050 and 1150, which correspond to the Little Springs volcanic event and immediate post-event ceremonial activities.

Along with the sherd rocks, other unique volcanic rocks were present. These specific rocks differ from the more common volcanic rock in this region. They are lightweight iron-rich basalts that were highly aerated. Along with the sherd rocks, these lighter volcanic rocks had to have been brought into this site from the lava flow.

Research has shown that there are many similarities between the responses to the Sunset Crater and Little Springs volcanic events in terms of powerful religious leaders interacting with active volcanoes and creating ceremonial objects. This is a result of Southern Paiute Puha'gants applying the TEK gained at Sunset Crater to Little Springs. Southern Paiute Puha'gants used their knowledge base and tried a new approach to the ceremonial activity. Instead of corn, Southern Paiute religious leaders used whole pots. This change is reflective in that the Paiute response to the Little Springs eruption was a well-thought-out and well-planned cultural response to the Earth being reborn. This cultural knowledge was the intellectual foundation for the Southern Paiute cultural responses to the Little Springs volcanic event.

Post-Eruption: The Initial Southern Paiute Response

There is extensive evidence of post-eruption Southern Paiute construction activities throughout the entire lava flow. Currently, it is estimated that there are more than eight kilometers of trails across the basalt lava flows. In addition to having brought in burden-basket load after burden-basket load of cinders, Indian people constructed hundreds of low wall and generally round structures along these trails.

It is believed that constructing the trails and structures on the lava flow was very labor-intensive and took a long period of time to build. It is important to remember that 'A'ā (rugged Hawaiian-style) basalt lava flows like Little Springs are not smooth but are instead characterized by sharp vertical edges with deep cracks between. There are very few loose basalt stones and almost no cinders available anywhere on the flows. While the surrounding volcanic mountains that define the two valleys do have cinders, these are greatly intermixed with dirt from the forests that have grown upon them. The only abundant source of cinders is the original pyroclastic mountain that was the source of the lava flow. Constructing both trails and structures was thus a significant engineering challenge.

The trails being made from imported cinders occasions an analysis of the level of efforts required to move these cinders to the lava flows. We can begin with the means of transporting the cinders. The only known carrying device available at the time is the Southern Paiute burden basket (Fig. 4). Basket loads would have been transported while held with a tumpline or positioned across the forehead of a man or woman. A typical Southern Paiute burden basket is 76 centimeters tall, 71 centimeters wide in diameter at the top and pointed at the bottom. We estimate that this would hold about 2.8 bushels, or about 100 liters. The known trails have been estimated at eight kilometers. If we look at how many cinders it would take to make a meter of these trails, we estimate it would take three burden-basket loads per meter. Eight kilometers of trail equals 8000 meters. When multiplied by three loads per meter this suggests that about 24,000 round trips were needed to make these trails.

Fig. 4 Southern Paiute burden basket. (Photo: Kathleen Van Vlack & Richard Stoffle)

Over 100 structures were built directly on the basalt lava flow and thus would have involved an unknown number of man-hours to build the walls from loose basalt blocks, most of which would have to have been carried in from elsewhere on the lava flow. The rockfill in the structures would have involved the same kinds of efforts as were needed for the construction of the trails. We estimate that about 1200 liters of fill would be needed per structure at a minimum. It thus would take about 12 burden-basket loads of cinders to make each structure usable. We estimate that 100 structures would involve about at least 1200 round-trip basket loads.

When the basket loads for the construction of both the trails and the associated structures are added together, it totals an estimated 25,200 round trips. In addition to the labor needed to construct the walls of the various structures, it can be assumed that the walls were complemented with sidewalls and conical roofs made from plants. More labor was needed to plan and supervise this elaborate series of trails and low walled structures on the two lava flows. All persons involved in the construction would have been fed and attended to by a support community. The constructions on the Little Springs Lava Flow were a major undertaking, but perhaps of more import for the cultural assessment of the area is where the idea for such construction came from and why this cultural innovation was made at all.

The only published theory for interpreting the trails and low walled structures on the two lava flows is by Mark Elson and Michael Ort (Elson & Ort, 2006). Their theory is that after the lava flows cooled, Indian people came under attack by others. Their response to this attack was to build a place to hide in the lava flows. Afterward, when they were threatened by attack, they could retreat to the lava flows for protection.

The Southern Paiutes have rejected Elson and Ort's theory and provided their own alternative theory. Southern Paiutes stipulate that this area had been used since time immemorial for ceremonies and healing. The Little Springs event both validated that previous cultural interpretation and added new Puha to the area. Indian people who experienced the Little Springs Lava Flow interpreted it as a gift of Mother Earth. The eruption and subsequent flow involved the emergence of Puha from the Earth in this special place. The eruption was a symbolic gift validating the power and cultural importance of this landscape. The only possible response for the people experiencing this gift was to lay the foundation for directly interacting with the volcano and its lava flow. An enormous amount of effort, under the guidance of religious leaders, was thus expended to provide access to special areas on the flow and structures so that religious people were able to spend time interacting with the flows and their Puha. It is assumed that the construction plans and their implementation occurred soon after the eruption event and took about a generation of concerted efforts. It is also assumed that this was a project that involved all of the Southern Paiute people who would attend to the ceremonial needs of this important spiritual area, as they did with other such areas.

On top of the lava flows located along these trails and throughout the flows are about 100 structures. This number is a very conservative estimate. Recent archeological survey work suggests that there may be as many as 250 structures on the lava flow (Hintzman personal communication). The typical structure is approximately four meters (~13 feet) across. Some are smaller and others much larger. Most structures today have evidence that their bottoms were filled with cinders and perhaps dirt. We estimate that a layer of fill up to 0.3 meters thick was needed to make the structures usable.

Post-Eruption Ceremonies: Little Springs Pilgrimage

After the volcanic event ended, Southern Paiute volcano Puha'gants continued to use this area for ceremony. The lava flow became a pilgrimage destination place for those seeking Puha (the energy force found in all elements of the universe) from the volcano and the lava flow. They followed a pilgrimage trail that led them to various shrines for prayers and rituals to prepare them for interaction with the lava flow and receive its knowledge and power (Van Vlack, 2012a, 2012b, Van Vlack et al., 2013). When the Puha'gants reached the northern lobe of the lava flow, they visited the hot spring. They used the hot spring for ritual cleansing and purification, which is a necessary step in the pilgrimage process (Van Vlack, 2012a, 2012b). The Puha'gants needed to have themselves cleaned of impurities and sickness, which could tamper with the ceremonial activities that occurred on top of the flow. During the post-volcanic-event pilgrimage, the shamans traveled along the trail that led them to an area known as Coyote's House.

Hot Spring at the Northern Lobe of the Little Springs Lava Flow: *Unuvats*

As pilgrims traveled south from Paiute Cave toward Mount Trumbull, the valley becomes bounded by high volcanic mountains and flows. These types of constrictions are important physical attributes to Southern Paiute pilgrimage trails. In Southern Paiute culture, narrow or constricted spaces influence cultural meaning and affect the movement of natural elements like wind and water. Pilgrimage trails such as this one to Little Springs pass through or near these narrow spaces because these are areas where Puha converges and collects, like how water will pool in constricted places. As a trail passes through these types of locations, a pilgrim can experience and draw upon the power of the area as he or she progresses on the journey.

As pilgrims traveled southward, they reached the northern end of the Little Springs Lava Flow to the hot springs. According to tribal representatives, the name for this spring in Southern Paiute is *Unuvats*, which means "volcanic rock spring." The Northern Lobe hot spring was formed from local volcanic activity in the area. When the Little Springs Volcano erupted, the flowing lava altered the area's hydrology. As the lava moved across the landscape, it followed existing surface drainages, blocking them and forcing the creation of new surface water flow patterns. The lava flows are very permeable, allowing for the water to be stored in naturally occurring reservoirs and recharge the water table.

The Puha that emerges from a place can be used to heal individuals and small groups (Miller, 1983). The power of the place is often supplemented by the presence of medicinal and ceremonial plants, minerals, viewscapes, and another natural phenomenon. These elements and artifacts are made more powerful by their proximity to powerful places.

Before entering a hot spring, Indian people would speak to the spirit of the spring, introducing themselves and explaining why they were visiting (Stoffle et al., 1997). Southern Paiute elders have stated that hot springs were also used by shamans for ritual purification before visiting sacred caves, valleys, or other spiritual locations. Such purification was necessary to prepare the mind and body for safe and proper interaction with spiritual beings (Stoffle et al., 1997; Van Vlack, 2012a, 2012b).

There are remnants of numerous stone structures located at the base of the flow near the spring. These structures are believed to be associated with ceremonial activity and purification at the hot spring for Puha'gants or shamans before visiting the lava flow. The Puha'gants most likely used structures in correspondence with the hot spring, and this was an intense period of prayer and preparation.

The Northern Lobe of the Little Springs Lava Flow

The Northern Lobe of the Little Spring Lava Flow (Fig. 5) is found at approximately 2018 meters (6620 feet) in elevation and is situated in a valley between Mount Trumbull to the northeast and Mount Logan to the southwest. The previously discussed hot spring is found at the base along the flow's northern edge. The pyroclastic cone responsible for creating the Little Springs Lava Flow is located approximately 1.92 kilometers (1.20 miles) to the south of the Northern Lobe's northern edge.

Puha'gants started their ceremonial activities at the Northern Lobe by first purifying themselves at the hot spring. From there they climbed up onto the lava flow (Fig. 6). Along the northern edge, there are two primary trails. The trail closest to the hot spring leads directly to a hornito. It is believed that Southern Paiute religious specialists used this trail to

Fig. 5 The Little Spring Lava Flow with structures in the foreground. (Photo: Kathleen Van Vlack)

interact with the lava flowing up from the hornito to create the sherd rocks. The second trail is believed to be associated with the pilgrimage to the center of the Northern Lobe to an area known as Coyote's House.

The trail network across the Northern Lobe involved extensive construction following the volcanic eruption. Indian people began to build a series of trails and ceremonial structures on the lava bed. Construction involved bringing in cinders from the Little Springs Volcano itself in burden baskets in approximately 25,200 round trips. This estimate accounts for eight kilometers (4.97 miles) of trails and a conservative estimate of 100 structures. UofA ethnographers and Southern Paiute representatives noted at least ten distinct circular structures ranging from three to six meters wide. Cinders were used in wall construction and floor fill. Indian people also used the cinders to fill in the hornito bottom to create a stable floor base, so religious specialists could be inside the hornito during post-eruption ceremonial activity.

Fig. 6 Trail leading to the Northern Lobe. (Photo: Kathleen Van Vlack)

Coyote's House

The other trail leads to an area on the lava flow referred to as Coyote's House (Fig. 7) or Coyote Pocket. This feature is a large volcanic outcrop that has a subsurface cave and unique vegetation. Along the trail to this location, many pieces of pottery were found. Additionally, numerous naturally occurring water catchments are found in this part of the lava flow. A large bowl fragment was found sitting above one of the largest water catchments at this site.

During ethnographic interviews with Southern Paiute tribal members, they discussed how the Northern Lobe was a ceremonial destination place used by special types of medicine men or shamans. In order to understand how and why Southern Paiute shamans came to the Little Springs area for ceremony and Puha acquisition, it is necessary to examine the relationship between religious practitioners and Puha. Shamans are predisposed to have certain kinds of Puha, which they can use to accomplish certain

Fig. 7 Coyote's House. (Photo: Kathleen Van Vlack)

things that may or may not benefit Southern Paiute society. Shamans gain access to additional Puha by visiting special places and through the acquisition of spirit helpers. Sometimes, they use their Puha to bring people, communities, or the world back into balance. During these times, the shamans are portals through which their Puha and the Puha of their spirit helpers combine during the ceremony to achieve a cure.

The Puha'gants left offerings at these water catchments, and they prayed and sought the proper permissions to enter into Coyote's House. Inside this unique geologic structure is a large cave, which Puha'gants would enter for ceremony. This area could have served as a portal where the Puha'gants left this dimension of the universe and entered into another one to acquire Puha associated with volcanism. The power and knowledge the Puha'gants gained during this portion of the pilgrimage were used in ceremonies in their home communities or at other ceremonial areas, such as places in the Grand Canyon.

Discussion: Navigating the Epistemological Divide

At the crux of this environmental communication debate between archeologists and Southern Paiutes is the question of who can speak about the past and who is considered the authority or expert. As we can see from the analysis, the Southern Paiute people understand the Little Springs Lava Flow to be intimately connected with their history and their cultural heritage. They have oral histories linking the topography of the landscape to their history and their beliefs. The origin stories of the Southern Paiutes stand in conflict with the findings and interpretation made by archeologists in the same area. Great Basin and Colorado Plateau archeologists argue that Southern Paiutes are recent arrivals in the area, and either they moved into a land emptied by the migration of its former inhabitants, or those inhabitants died out where they had lived for thousands of years. These arguments suggest that the Paiute were a replacement culture characterized by nomadic wandering and hunting and gathering. Either argument can be categorized as a component of the *Numic Spread Hypothesis* (Lamb, 1958), which archeologists use to distance and disconnect Paiute people from their homelands and cultural resources. Archeologists who conducted research in and around the Little Springs Lava Flow have brought forth this hypothesis in their analysis, and thus they have attempted to shut the Paiutes out of the story.

The Kaibab Band of Paiute Indians has formally rejected all aspects of the *Numic Spread Theory* since the theory's rise to prominence in Great Basin and Colorado Plateau archeology. The tribe, along with the other seven Southern Paiute tribes, has stipulated that the Southern Paiute people have lived in their traditional lands from *time immemorial* (legal term), or since the time when they were placed on these lands by the *Creator* (Stoffle et al., 2004; Southern Paiute Advisory Committee, 2011). They have also stipulated that they are the ones who used and interacted with the Little Springs Lava Flow during and after the eruption.

As a heritage issue and an environmental communication issue, how does one reconcile the stark differences in interpretation of how the Little Springs Lava Flow has been used since its eruption? The Kaibab Paiutes, like many other indigenous groups, have the right to tell their own story and the right to determine how their past should be investigated and shared with the public. This issue is the result of two vastly different epistemologies, and often the differences cannot be dealt with in a manner that benefits both sides. One solution which has proven to be successful throughout the U.S. National Park Service is to move away from the desire for only one ultimate story/one interpretation and allow for multiple voices in an effort to settle the environmental communication issue and cultural heritage dispute. This allows the Kaibab Paiutes to share their perspectives and cultural understandings of Little Springs Lava Flow, and it would also allow the archeologists to present their interpretations as an alternative way of understanding. When tribes like the Kaibab Paiutes are included as partners in the decision-making process, they will bring forward their cultural knowledge as a valuable tool in heritage management and heritage conservation.

Acknowledgments Dr. Van Vlack wishes to express her sincere gratitude to all of the Southern Paiute people who participated in the study. She expresses her heartfelt thanks to them for taking time away from their families and busy schedules to share their knowledge of traditional life and cultural resources in the Arizona Strip. Without the collaboration and participation of Southern Paiute elders and experts, this study could not have been done. Dr. Van Vlack also wants to thank the Bureau of Land Management Arizona Strip Field Office and Grand Canyon–Parashant National Monument for providing the funding and agency support for this research.

References

Elson, M., Ort, M., Hesse, J., & Duffield, W. (2002). Lava, Corn, and Ritual in the Northern Southwest. *American Antiquity, 67*(1), 119–135.

Elson, M. D., & Ort, M. H. (2006). *The Little Springs volcanology and archaeology project, Grand Canyon–Parashant National Monument, Arizona, investigators final report.* Western National Parks Association.

Goldman, A. (1999). *Knowledge in a social world.* Oxford.

Hanson, A., & Robert, C. (2015). *The Routledge Handbook of Environmental Communication.* New York, Routledge.

Hintzman, M. (2012). *Take me to Lava Town: The archaeology of the Little Springs Lava Flow.* Master's thesis, Flagstaff, Arizona, Northern Arizona University.

Joosse, S., Powell, S., Bergeå, H., Böhm, S., Calderón, C., Caselunghe, E., Fischer, A., Grubbström, A., Hallgren, L., Holmgren, S., Löf, A., Nordström Källström, H., Raitio, K., Senecah, S., Söderlund Kanarp, C., von Essen, E., Westberg, L., & Westin, M. (2020). Critical, engaged and change-oriented scholarship in environmental communication: Six methodological dilemmas to think with. *Environmental Communication, 14*(6), 758–771. https://doi.org/10.1080/17524032.2020.1725588

Kelly, I. (1939). *Southern Paiute Shamanism*. Berkley: University of California Press.

Lamb, S. (1958). Linguistic prehistory in the Great Basin. *International Journal of American Linguistics, 25*, 95–100.

Liljeblad, S. (1986). Oral tradition: Content and style of verbal arts. In W. L. d'Azevedo (Ed.), *Handbook of North American Indians, vol. 11: Great basin* (pp. 641–659). Smithsonian Institution.

Miller, J. (1983). Basin religion and theology: A comparative study of power (puha). *Journal of California and Great Basin anthropology, 5*(1,2), 66–86.

Milstein, T., & Kroløkke, C. (2012). Transcorporeal tourism: Whales, fetuses, and the rupturing and reinscribing of cultural constraints. *Environmental Communication, 6*(1), 82–100. https://doi.org/10.1080/17524032.2011.642079

Mohanty, C. T. (1984). Under western eyes: Feminist scholarship and colonial discourses. *Boundary, 2*(12/13), 333–358. https://doi.org/10.2307/302821

Ort, M. H., Elson, M. D., Anderson, K. C., Duffield, W. A., Hooten, J. A., Champion, D. E., & Waring, G. (2008a). Effects of scoria-cone eruptions upon nearby human communities. *Geological Society of America Bulletin, 120*(3/4), 476–486.

Ort, M. H., Elson, M. D., Anderson, K. C., Duffield, W. A., & Samples, T. L. (2008b). Variable effects of cinder-cone eruptions on prehistoric agrarian human populations in the American Southwest. *Journal of Volcanology and Geothermal Research, 176*(3), 363–376.

Rappaport, R. (1999). *Ritual and religion in the making of humanity*. Cambridge University Press.

Rhode, D., & Louderback, L. (2007). Dietary plant use in the Bonneville Basin during the terminal Pleistocene/Early Holocene transition. In K. Graf & D. Schmitt (Eds.), *Paleoindian or paleoarchaic? Great Basin human ecology at the Pleistocene–Holocene transition* (pp. 231–247). University of Utah Press.

Smith, L. (2012). *Decolonizing methodologies: Research and indigenous people*. Zed Books Ltd..

Southern Paiute Advisory Committee (SPAC). (2011). Southern Paiute Ethnographic Study for the Lake Powell Pipeline EIS: Report Regarding Study Plan 23-Ethnographic Resources, for Kaibab Band of Paiute Indians, San Juan Southern Paiute Tribe, and Paiute Indian Tribe of Utah, for the Lake Powell Hydroelectric System. Report Prepared for Utah Division of Water Resources. Submitted on Behalf of the Kaibab Band of Paiute Indians, Paiute Indian Tribe of Utah, and the San Juan Southern Paiute Tribe. Bureau of Applied Research in Anthropology, University of Arizona, Tucson AZ.

Stoffle, R., Arnold, R., Frank, M., Cornelius, B., Miller, L., Charles, J., Kane, G., Ruuska, A., & Van Vlack, K. (2015). Ethnology of volcanoes: Quali-signs and the cultural centrality of self-voiced places. In L. Scheiber & N. Zedeno (Eds.),

Engineering mountain landscapes: An anthropology of social investment (pp. 99–111). University of Utah Press.
Stoffle, R., Chmara-Huff, F., Van Vlack, K., & Toupal, R. (2004). *Puha flows from it: The Spring Mountains cultural landscape study.* Bureau of Applied Research in Anthropology, University of Arizona.
Stoffle, R., Halmo, D., Olmstead, J., & Dobyns, H. (1994). *Uatah (fields where we plant all the time): Shivwits Southern Paiute water use along Tunakwint, the Santa Clara River.* University of Michigan.
Stoffle, R., Van Vlack, K., Carroll, A., Chmara-Huff, F., & Martinez, A. (2005). *Yanawant: Paiute places and landscapes on the Arizona Strip—Volume one of the Arizona Strip landscapes and place name study.* Bureau of Applied Research in Anthropology, University of Arizona.
Stoffle, R. W., Halmo, D., & Austin, D. (1997). Cultural landscapes and traditional cultural properties: A Southern Paiute view of the Grand Canyon and Colorado River. *American Indian Quarterly, 21*(2), 229–250.
Toupal, R., & Stoffle, R. (2004). *Traditional Resource Use of the Flagstaff Area Monuments.* Prepared for the National Park Service Cooperative Agreement Number 1443CA1250-96-006. Tucson, AZ: Bureau of Applied Research in Anthropology, The University of Arizona.
Van Vlack, K., Stoffle, R., Pickering, E., Brooks, K., & Delfs, J. (2013). *Unav-Nuquaint: Little Springs Lava Flow ethnographic investigation. Prepared for Grand Canyon Parashant National Monument.* Bureau of Applied Research in Anthropology, University of Arizona.
Van Vlack, K. A. (2012a). *Puaxant tuvip: Powerlands Southern Paiute cultural landscapes and pilgrimage trails* (Order No. 3505682). Available from Publicly Available Content Database. (1013831696). https://lib-proxy.sunywcc.edu/login?url=https://lib-proxy.sunywcc.edu:2077/dissertations-theses/puaxant-tuvip-powerlands-southern-paiute-cultural/docview/1013831696/se-2?accountid=14174
Van Vlack, K. A. (2012b). Southern Paiute pilgrimage and relationship formation. *Ethnology, 51*(2), 129–140.

Arsenic Fields: Community Understandings of Risk, Place, and Landscape

Annelie Sjölander-Lindqvist

Introduction

This chapter is a study of the communication of environmental risk associated with the planned removal of arsenic from the copper mining fields in Riddarhyttan, central Sweden. The study, carried out in 2016–17, employed ethnographic interviews with community residents and representatives from the local, regional, and national governments, as well as walk-and-talk sessions near contaminated areas to learn about the value of these places to local people. The focus is on the local responses to the toxics and the understandings and conceptualizations of the local community: How did the local residents understand the toxic contamination? What are the social and cultural implications of the presence of toxics? Did the contamination and planned remediation affect the informants'

A. Sjölander-Lindqvist (✉)
School of Global Studies & Gothenburg Research Institute, University of Gothenburg, Göteborg, Sweden
e-mail: annelie.sjolander-lindqvist@gu.se

A. Sjölander-Lindqvist et al. (eds.), *Anthropological Perspectives on Environmental Communication*, Palgrave Studies in Anthropology of Sustainability, https://doi.org/10.1007/978-3-030-78040-1_3

perceptions and views of the landscape, place, and locality, that is to say, their local environment? How did the informants navigate between contradictory agency messages regarding the risk of residing in a contaminated local environment? As we will see, the answers to these questions are complex and beyond the scope of technology and straightforward communication. In addressing the ways in which the planned clean-up and the communication of the risks posed by the toxics in the local environment were understood, it will be clear how community responses are culturally informed, historically conditioned, and subject to technological and scientific uncertainty.

To borrow from Depoe (2007), this study locates itself in the field of environmental communication through its focus and intention to address how people "locate or situate themselves within a geographic space or cultural scene … that has meaning or significance for their identities and relationships" (p. 3). Depoe's argument that we ought to know more about how "humans use symbolic resources to define and modify their surroundings, their environments" lends a perspective familiar to social anthropologists. For the anthropologist interested in landscape and what makes place a "place," the issues of lived reality and symbolic meaning are crucial, and this is where anthropology meets the field of environmental communication. The perceptions and values of the local community, including the landscape and particular places where people live and work, are important entry points, since the tangible conditions in a local environment, and the associated values of its residents, mirror the understandings and experiences acquired from lived, everyday involvement in the world (Ingold, 1993).

In addition to the understanding that the landscape is a holder of symbolic meaning and a juncture where the past is in the present and the present is in anticipation of what is to come, the anthropological gaze is perhaps even more important in cases where technological projects may violate the basis for a community's values (Sjölander-Lindqvist, 2004). Such projects are usually embedded in a setting of information sharing and both formal and informal communication with the intention to create awareness and readiness for action among the receivers of the information, who may be the local community or other stakeholders (Argenti, 2002), and to legitimize interventions (Renn and Levine, 1991). The ethnographic tradition of describing and understanding "the multiplicity of complex conceptual structures" (Geertz, 1973, p. 9) can here serve as a methodological and analytical compass. Arguably, one important task of anthropology is to examine practices "in work" and focus on "the conditions that create and sustain them and the kinds of relations and subjects they produce" (Shore et al., 2011, p. 213; cf. Geertz, 1973).

The study therefore draws inspiration from Geertz (1973), who held that "the thing to ask is what [...] is getting said" and what the "import is" of the occurrence of the realities explored when entering the field of our investigations (p. 10). For reasons derived from theory and previous studies of similar cases (Sjölander-Lindqvist, 2004), this study's exploration of the communication of environmental risk and the associated understandings was approached as a bounded setting with unique tangible and associative circumstances (yet related to a wider scope through the responsibility of governments to protect the environment and human health, which in terms of polluted land finds expression in the widespread remediation and restoration of contaminated sites). This involved speaking with people about their way of perceiving the local environment and their ideas and understandings of the risk communication process. The research was carried out by the author and an assistant who assisted with notetaking during walk-and-talk sessions and transcription of interviews undertaken with former mine workers and their partners, homeowners and tenants living next to or on contaminated sites, and representatives of agencies at the local, regional, and national level. The collected data was reviewed and analyzed from the perspective of field-specific perceptions, that is, local life as lived and interpreted by the informants and the policy work as understood and interpreted by the concerned agencies.

As part of a larger research project that sought to explore the ways in which Swedish government actors approach and communicate risk, this study provides insights into how communication on environmental risk involves a melee of actors who each hold unique ways of conceptualizing and framing everyday involvement in the world. Actors' responsiveness to risk and uncertainty is a crucial area of study (Power, 2007), one with relevance for an anthropological study of the communication of environmental risk. The need for mitigation actions to decrease or eliminate emergent effects of hazards in the immediate environments of life—be they chemical agents, pollutants that provide insult to air, water, or soil, or the presence of biological agents (which in the current time of the COVID-19 pandemic has evolved as a highly acute risk to human life)—underscores the importance of identifying the ways people understand and negotiate the risks associated with living in toxic environments. In the anthropological tradition of studying risk and risk communication, the Riddarhyttan case is an illuminating example of how perception and communication is embedded in the social and cultural reality of local communities (e.g., Boholm, 2015; de França Doria et al., 2009; Sjölander-Lindqvist, 2004; Stoffle & Arnold, 2003).

Contaminated Places and Communities

Over the years, we have repeatedly witnessed cases of toxic contamination where local communities have confronted the risks associated with dangerous contamination of soil, groundwater, and surface water. Inventories show, for example, that around 1,170,000 potentially contaminated sites have been identified across 27 European countries, which at the time of the European Soil Data Centre report was estimated to be about 45% of the total number of possible sites (van Liedekerke et al., 2014). As of today, inventories have identified approximately 85,000 potentially contaminated sites throughout Sweden.

A well-known and catastrophic case within the field of contaminated land is the dumping of nearly 21,000 tons of chemical waste in the 1940s and the 1950s in the former Love Canal in the United States. The Love Canal disaster placed toxic contamination into the consciousness of the American people when bare-footed children had their feet burnt and black sludge filtered into gardens and leaked into cellars (Edelstein & Wandersman, 1987; Fitchen, 1989; Gibbs, 1998). Another example is the leakage of an acrylamide-based sealant and subsequent contamination of groundwater, surface water, and soil during the building of a railway tunnel, which became a huge environmental scandal in Sweden in the late 1990s (Sjölander-Lindqvist, 2004). In the United States, a class-action lawsuit against Pacific Gas and Electric over hexavalent-chromium-contaminated groundwater became a world-famous case when it was featured in the film *Erin Brockovich*, starring Julia Roberts, in 2000. The oil exploration and production activities from the 1950s onward in the Niger Delta region of Nigeria is another compelling case where the occurrence of oil in the landscape continues to have significant environmental, social, and economic impacts (Zabbey et al., 2017).

Even if these cases, among many others, have occurred in different times, in different geographical locations, with varying short- and long-term effects, and subject to legal contestations, they exhibit unity in the way that human-technological intervention in the landscape can cause ecological, economic, and not least social consequences for society, local communities, and their environments. In such cases, communication is likely to become a fragmented field when expert knowledge and advice aimed at the protection of environmental and human health meet the concerns of the local communities exposed to the contaminants and their diversified conceptualizations of the toxic—be it an oil spill, hexavalent chromium, acrylamide, arsenic, or any other hazard (Lidskog, 2011; Sjölander-Lindqvist, 2004).

The Contaminated Riddarhyttan Copper Fields

Riddarhyttan, home to around 400 people, is located in central Sweden. Mineral prospecting and extraction have a millennia-long history in the area and have had an influential role in the global production of copper (Bindler & Rydberg, 2016). The earliest evidence of iron production dates back more than 2000 years, and there is also vague evidence of metal production from the eighth century CE (Karlsson et al., 2015). Despite their importance for societal development, the remains of these earlier industrial and extractive activities may cause a potential risk to the environment and to people. Due to the dispersal of hazardous substances in soils, sediment, groundwater, and surface water, these sites should undergo treatment to remove polluting substances. This was also the case for Riddarhyttan, where two copper worksites (the "Old Copper Works" and the "New Copper Works") have been screened for toxics (arsenic, lead, mercury, cobalt, and copper). Located in the historic heartland of the Swedish mining, metallurgy, and industry, Riddarhyttan can thus be added as another case of environmental risks, both in terms of local impacts and in terms of the complex process of communication.

Following environmental regulation and the implementation of the Environmental Quality Objective "A Non-Toxic Environment," an initial inventory was carried out in the later months of 2014. The results of the investigation indicated high levels of arsenic in the area of the New Copper Works. In March 2015, the Swedish Environmental Protection Agency decided to finance further examinations of the contaminated land with the Swedish Geotechnical Institute (SGI), as the main organizer ("huvudman") of the ensuing preparatory actions and post-treatment since the local municipality said they had small opportunities to lead a complex, technology-driven, and expensive remediation process. The SGI is an expert agency that works for a safe, efficient, and sustainable construction and sustainable use of land and natural resources.

The two copper works were put into operation in the early 1800s, but shut down in 1873 due to the low concentration of copper in the ore, and many remains of the industry are still visible today. Bare slag heaps and fields, dams, a water pipe, a few houses of stone and timber, yellow/orange-colored stones in the excavation slag fields, and up to 400 mine pits of various sizes serve as material reminders of the copper works (see Figs. 1 and 2). In addition to these historical remains of the copper industry, the Bäckefors Mine—in operation until 1979—is another important site in the area, both visually and experientially.

Since the copper mining fields are a historic environment, the examinations had to include an archeological component. The Department for the Cultural Environment at the County Administrative Board was therefore involved in the ensuing preparatory work. Concomitant with the archeological examination of the copper mining fields, the SGI, the main organizer of the remediation, dealt with the issue of fencing off or excavating the fields to hinder further community exposure and damage. This resulted in the decision to fence off parts of the fields and putting up warning signs

Fig. 1 Slag fields, Old Copper Works. (Photo: Annelie Sjölander-Lindqvist)

Fig. 2 Mining remains. (Photo: Annelie Sjölander-Lindqvist)

next to the arsenic-contaminated copper fields to inform people of the risk of entering the fields contaminated with the toxic residue of copper extraction.

In the fall of 2015, the SGI decided to complement the initial toxic analysis using another, more expensive method of assessment. In January 2016, the results of the new analysis were made official. The analysis generally indicated considerably lower levels of human health risks. In consequence, the organizer considered altering the remediation plan. On the other hand, one component of the new analysis, namely the leaching tests, confirmed the initial risk level, and so the SGI still had to decide whether the arsenic remains in the fields were of such high levels that they could jeopardize peoples' health. According to an interview with an SGI representative, they finally determined that "one could spend quite a lot of time on the fields, and one would have to eat a substantial amount of arsenic-polluted dirt for it to be a health risk." Based largely on the results of the second toxic analysis, which included the new leaching tests, the SGI therefore decided to postpone remediation and make new inventories including more substances in the evaluation. The warning signs that had been posted next to the contaminated sites were removed, and later re-posted. Based on the interviews with the agency representatives, however, none of them knew why the signs were re-posted, or even by whom. The agency informed the property owners that the results of the new tests did not indicate a risk to human health. As the agency stated, there was a low risk of human body absorption, and the agency decided that visitors and residents could enter the fields without risk to human health.

Communication of Environmental Risk

The use and exploitation of land and natural assets have in many cases put the environment at risk, affecting both biophysical and social environments. The legacies of the industrial past and the continuous extraction of natural resources have a salient influence not only on contemporary land use and management but also, as we will see, on community understandings and risk perceptions. To avoid public misperceptions and inadequate comprehension of technological interventions and ensure that public values and priorities are included in project management, government administrations and industry employ risk communication. This field of communication arose largely from "the disconnect between public risk perception and expert risk assessment" (Boholm, 2015, p. 154) with the aim of changing the public's attitudes (Fischoff, 1995). Whereas risk communication in the late 1960s and 1970s was concerned with unilaterally providing the public with information, communication on risk now involves procedures that favor public participation and dialogue (Boholm, 2015). This move toward two-way communication continues to be far from clear-cut. Outcomes and probabilities of certain actions aimed at the reduction of risk are still contradictory and muddled by discursive disagreement regarding proper scientific modeling to establish "truth," and political and economic considerations disputing priorities and ends

(Boholm, 2015). Consequently, the best way of communicating risk is still being discussed in public as well as in scholarly debate.

Besides the fact that individual and community understandings of risk are contextual, relational, and rooted in cultural values and meanings (e.g., Adger et al., 2013; Boholm, 2015; Boholm & Corvellec, 2011; Cutter et al., 2008; Mairal Buil, 2005; Rappaport, 1996; Sjölander-Lindqvist, 2004), local responses are also dependent on the receiver's trust in the sender (e.g., Steelman et al., 2015) and whether the two parties in the communication process share the code of the language used (Zimmerman, 1987). For example, if the language is too technical, the receiver might be unable to understand or unwilling to listen to the message, leaving the receiver confused or emotionally upset (ibid.). Likewise, as in the present case: if the sender's risk signal pends between different scenarios and messages, the receiver of the signal is left to their or own interpretation and judgment.

The muddled character of risk communication is quite clear in the Riddarhyttan case, and the social amplification of the risk framework (Kasperson et al., 1988; Renn, 1991; Pidgeon et al., 2003) can be of help here. This framework points to how "risk signals" can be distorted due to attenuated or amplified risks when a transmitter of information either exaggerates or downplays the risk. Whereas communication indeed involves a sender of information and a receiver of the messages sent, the social and cultural conditions and contexts of the given communication must also be acknowledged (Boholm, 2015). Here, we can borrow from the social amplification framework, but with a reverse twist, in the sense that it is the receiver, not the sender, who may either amplify or downplay the possible risk. This prompts us to explore and understand the reactions and feelings arising when oscillating signals of risk confront a local community and this forms the basis for risk conceptualization in the post-treatment of contaminated places.

Local Community Perspectives

For the Riddarhyttan local community, the presence of toxics is well known. Driving into the area, one sees many remnants that explicitly show that this is a place of mining. Slag fields, slag heaps, dross mounds, mining pits, mining buildings, and infrastructure are clearly visible in the landscape (see Fig. 3). These material remains are a reminder of how interventions in the biophysical environment affects the natural and built

Fig. 3 Dross mound. (Photo: Annelie Sjölander-Lindqvist)

environments not only in terms of visibility but also in terms of social and cultural meanings and impacts on people who face these facilities or remains of the construction (see, e.g., Edelstein, 1988; Fitchen, 1989; Mairal Buil, 2005; Sjölander-Lindqvist, 2004). As was demonstrated in the interviews and the walk-and-talk sessions, the mining industry and history is a vital part of the community inasmuch that the term "mining community" seems a relevant conceptualization here. Similarly to studies on communities in the proximity of nuclear power stations, a greater understanding of the factors that affect and "drive" the building of meaning can in fact offer important insights into how different actors construct and reflect on risk (cf. Venables et al., 2012).

Besides the obvious material remains of the mining industry, the meanings associated with the mining industry also include an immaterial dimension, reflecting the life both below and above ground. These memories are more associated with the more recent mining activities of the Bäckefors Mine, but also reflect the societal role of the mining industry. As one of the informants explained:

> this great sense of cohesion; we relied on one another underground, we spent time together in our spare time, dancing every weekend. This sense of belonging to this place, this community and the mine, being part of the history and the industrial era. Then, the emptiness, when the mine closed down. I don't think anyone would hesitate to go underground again.

In addition to the place where people live and work being historical, the local environment is described by the informants as amazing, as a "green" place with "fresh air," but with a "tough but hearty industrial mentality" that is "exciting" and "a good place to raise children." Even if there is shared knowledge regarding the toxicity, the informants downplay the toxic risk. "Everyone knows about the toxics," they say, and during the walk-and-talk sessions the informants do not hesitate to enter the slag fields. Signs next to the fields warned of the contamination, but the clearly strong and positive-laden sentiments about these places sometimes outweighed the risks. One informant said that the "contamination is part of this place" and the majority of the informants considered the toxics as a "natural" part of the local environment. In addition, the reagents used during the extraction process to purify the copper were said to be "natural" as a way to further stress the naturalness of the place. The strong sentiments regarding the local mining industry and the local community were also an important dimension of the community members' understanding of the risks associated with the toxics, which at least to some degree are present in the local environment. This way of describing the environment and the presence of toxics enhanced locals' sense of belonging to the place.

While the informants demonstrated a strong sense of pride and attachment to the place, they also discussed how they live and move around in the area to avoid the toxics. For example, they discussed how one should not eat fish from the lake and the waterway connecting the two copper works, since the toxics "make the fish see-through," nor should one swim in the lake: "When I moved here they told me not to bathe or fish in the lake!" One woman talked about the time when she was pregnant and said: "I decided not to worry since I didn't really know how risky it would be, to live here, pregnant. I never really decided to find out." In general, people who had been raised near the sites and had spent their childhoods playing on the slag fields, dross mounds and slag heaps had always known that "one shouldn't go to the copper fields when it was raining because of the release of sulfur from the slag." Likewise, the informants with younger children spoke of how they tried to inform their younger ones about the risks of spending time outside—"Don't go swimming in the lake!" At the same time, there were contradictory opinions. As one informant expressed: "if we can't move around in the area, we can't live."

These contradictory feelings toward the environment were shown through the relating of anecdotal evidence. For example, it was explained how the risks were exaggerated. Narratives focused on the experiences of community members, and of low risk—someone had a neighbor who ate fish from the lake and "he lived to a very old age." One informant said: "Look at me, I'm old. As a child I always played on or near the dross mounds. Can you see a difference? No. It's nothing dangerous."

Contradictory feelings toward the toxics were also evident among those informants that had not been born or raised in the area and had moved there. When friends come to visit them, they felt that they have to show that they have no real fears of the area being toxic.

> We just keep our distance; we can't take this seriously. We can't say 'This is dangerous! Don't eat this!' Cause then they would ask: 'But why do you live here, then?' It makes your decision seem really strange. … So, when visiting friends ask, I say, be careful of the mine pits and the arsenic, but at the same time, I try to be as vague as possible.

This contradictory anxiousness and uncertainty is also reflected among the families who have moved to the area when talking about everyday activities, such as growing your own vegetables or bathing in a small community pond that lies next to some smaller and larger water-filled pits. Struggling with the uncertainty regarding the degree of risk, one informant said, "the pond is probably not super-toxic, but in reality we have no idea." Moreover, "We grow vegetables, since we think the produce is good and healthy, but then, you think, perhaps the ground is full of toxics?" Another informant said, "I continue growing veggies. … I pretend everything is fine, but I really don't know." Since "we're still alive," it cannot be that much of a risk, the informant continued. It may be that, the word goes that the residue from the copper purification process actually "improves and cleanses the soil."

Risk Communication in Riddarhyttan

The risk communication that took place in Riddarhyttan came in three forms and contained contradictory risk messages: through letters to property owners, through posted warning signs next to the copper works and the adjoining slag fields, and through local and social media. In November 2014, the local news media stated that "residents [of Riddarhyttan] should not touch stones because there is a risk of dust from the mining waste" (Olsson, 2014, 5 Nov). A few weeks later, local news media reported that "residents can continue drinking water from their wells," (Sveriges Radio Västmanland, 2015, January 2) and in June 2015, the media stated that "It is not dangerous to walk around in the area, but if you touch stones and then put your fingers in your mouth, it can be life-threatening" (Högberg, 2015, June 17). The warning signs (later removed), posted soon after the initial tests, gave notice of a high level of arsenic contamination in the fields. The letters sent to the property owners living in proximity to the arsenic fields informed them that "visitors and residents can use the area without risk to human health," and that "the tests show that the arsenic is strongly bound to surface mining waste. The portion that can be absorbed by the human body is low."

Even if there were concrete forms of risk communication, the interviewed agency representatives at the local, regional, and national levels showed a striking uncertainty regarding the risk communication process. Discussing the risk communication process, the informants from the different government agencies considered it important to have a clear idea of who has the main communication responsibility and what to communicate. This seems to have never really been the case. As one of the agency representatives said, "it is difficult to communicate in such a pressing context of pervasive uncertainty as to whether there were any toxics at all in the area or if it posed a risk to humans." Another example is that none of the interviewed agency representatives seemed to know when the decision to remove the posted warning signs had been taken, or who had actually removed the signs after the results of the second toxic assessment had made it clear that there was a low level of risk to human health.

Following up on the uncertainty regarding the responsibility for communication, one of the local representatives of the municipality said that it was better if the regional authorities run the communication process, since "the residents take the County Administrative Board more seriously." What to communicate remained overall uncertain to the informants at the different government agencies, and one representative said: "What should we actually tell people? What is it that is dangerous? We just don't know."

Discussing risk communication, a national agency representative said that the agency never really "thought about how local people thought about this; we just proceeded from our own perspectives, how we should deal with the remediation." In retrospect, the representative said that the risk communication that had taken place "was made up; we thought this was a good way to do it," but regretted that it "did not support the building of trust." Some agency representatives at the local and regional levels said they took this approach because they believed that the local community and the people who lived in the midst of the contamination "probably" reasoned that there was no need to worry, since the toxics had been present "forever." The representative continued, downplaying the risk by saying, "It's everywhere, and no one has died from it yet."

At the same time, one of the agency representatives claimed that "communication must be given time; it's not something to be rushed, lest the communicators lose their control of the communication process." The representative reasoned that full transparency was desirable but debatable:

> How do you inspire and establish trust when you're reporting contradictory test results? What is it that makes one test more reliable than the previous test? I think it's easy to imagine that yet another test result might come, another assessment that renders yet another result.

Coming to grips with the dilemma that so visibly arose in the Riddarhyttan risk communication process, the national agency representative said:

> I don't think we should just say this and that and leave it to the receiver's interpretation; nor is the reverse particularly wise. I think we need to give it some more thought, how to actually communicate, think it over, what's the right way. The way we did it in this case, declaring a state of emergency, that's kind of like giving a red light; we lost it, we should have given it a

second thought, the choice of conceptualization is important, choosing 'warning' signals is a great risk. These double messages don't inspire trust.

The uncertainty that ensued following the results of the different screenings for arsenic contamination, the communicated degree of toxicity, and the associated presumed risks underlie the mixed community response. Residents said that "there's no information" and "what we know, we've learned ourselves." This could be summarized as confusion about whether there was any viable information at all. Local enquiries about who was actually responsible for the communication of risk and what was really going in terms of risk assessment and remediation plans were partly left unanswered. Facing this uncertainty, some informants expressed that it might simply be best "not to know," "what is not known doesn't exist," and that "things might get worse if we start messing around with the soil." The logic here was that people would still start worrying, perhaps for no reason. They said that many local people grow old, have good memories, and are of good health, and therefore "it can't be that risky or dangerous." Or, as one of the informants explained, "I know too little to worry." In addition, if the agencies decided to abandon the remediation plan, one of the interviewed residents said, "it can't be that bad."

There were also different responses depending on how long people had resided in the area. Those who have lived there since they were young or had worked in the mine appeared less worried and less interested in being informed. The "newcomers" on the other hand, were commonly less relaxed, even if they, like the others, were also inclined to describe the risk as rather low. After all, they said, the mine and the toxic remnants of the excavation industry are part of the local heritage, which contributes to defining the place.

The reactions to the communication concerning the arsenic contamination were also obscured by certain other circumstances in the community. Many of the local informants and municipality representatives returned to the issue of the motocross track: the dust caused by the bikes and the annoyance to the residents living close to the track. This, some said, eclipsed their worries regarding the risks of arsenic contamination. The two separate issues arose at the same time as the County Administrated Board distributed a letter to the households to inform them of the arsenic in the soil. The dust issue "exhausted us," they said, and they did not have any energy left to worry about the arsenic. The question of whether to worry or not was exacerbated by the vacillating information, adding to the state of uncertainty:

> first we receive a letter that our health is at risk, then we're informed that it isn't as dangerous as they thought, so what is it really? Bewildering. Well, it made me happy to hear that the soil wasn't too contaminated and that it wasn't as dangerous as they'd first said.

Landscape, Place, Risk, and Memories

The interviews and informal conversations during the walk-and-talk sessions where the informants showed and discussed what the slag fields and water-filled mine pits meant to the industry helped paint a richer picture of the area and the meaning the remnants had to them. Keen to show the remains of the industrial past and how they thought about the communication that had taken place, they provided a deeper level of insight into the meanings that are attached to the remains of the past. This past is a lived reality that encompasses and reproduces the foundation of the local community. The underground extraction of minerals laid the foundation for the livelihoods of the local community members, and both the prehistoric and more recent past provide important means for the formation of local identity. The narratives about the past and the lives linked to the mines and the local environment constitute a collective experience and memories, based on relationships to a mining space bestowed with meanings and values (Sjölander-Lindqvist, 2004; cf. Stoffle & Arnold, 2003).

This mythopoeic past adds to the cultural and historical value of the area and the notion of landscape as something produced by local practice. Recalling how the informants throughout have been engaged—sometimes bothered by the past and the contamination, sometimes not—in the community and have lived their lives reminds us how landscapes are produced, modified, and contested by people who engage with them (Ingold, 1993). This emphasizes the role of process and the aspect of time and temporality. This "dwelling" perspective proposes that the landscape is the setting for human activity, and as such, the landscape comprises the totality of practices and experiences by those who "dwell" (i.e., perform all those activities that humans perform when they live their lives) in it (Ingold, 1993). Additionally, the cultural heritage remains—taking material shape through yellow-colored stones, water pipes, heaps, pits, dross mounds, and slag fields—are visual reminders and anchors for the symbolic meanings attributed to the mining past. Here, a scenic aesthetic perspective of the landscape connects with the temporality of landscape (Inglis, 1977; Ingold, 1993). The results of this study signify how the local environment, with its landscape, wildlife, buildings, landmarks, and monuments, is an important and meaningful resource, as it contributes symbolically to collective identification (Halbwachs, 1992 (1952); Schama, 1995).

The narratives—located in the past and in memories of life as it has unfolded—remind us how the landscape has the capacity to reinforce the past's connections to the present. The community members continue to weave webs of meaning based in the past. Not only do these narratives contribute to the informants' sense of belonging to their place of residence, but their past and lived experiences also inform their understandings of the contaminated landscape. They have always known about this contamination; this knowledge and the past help them to make sense of the presence of the toxics in today's community. Comments such as "there are so many things that are risky" and "the mining industry dates long back and has given us this unique cultural landscape" reflect what Appadurai (1995) referred to as the "socialization of the landscape" and tell us how the locales and the landscape constitute are a life story.

The story of the Riddarhyttan community has evolved over time, created and known through shared experiences drawing on collective and meaning-shaping symbols. This emphasizes how places and landscapes are where experiences and actions are carried out, where thoughts and expressions are created and knowledge is produced (Casey, 1996). For the local community, to be in this contaminated landscape is to be part of a time-place life story (cf. Stoffle et al., 2013). During the plans to remediate, this became even truer for the community: the toxics are a natural part of the locale and the living collective memory. This hazard is understood within a social and cultural context, which influences how people respond to risks stemming from both unforeseen and foreseen hazardous activities and occurrences. Hence, the presence of the toxics, interpreted as potentially risky, is an eventuality intrinsically connected to the notion of experience, shown in statements such as "it's not dangerous," "people grow old," "I worry more about the contamination I was exposed to in the mines," and "despite the presence, we could have children."

Growing your own vegetables, going swimming in the community pond on hot summer days, telling your friends that they should be careful in the area without overly stressing the risk, reading and reflecting over agency signs about the areas with contaminated soil, and undertaking various risk-reducing strategies are all routine activities that contribute to the sense of locality. These daily activities are also, as discussed by Tilley (1994), "biographic encounters for individuals, recalling traces of past activities and previous events and the reading of signs" (p. 27) and performed in order to realize the social ideal of locality (Appadurai, 1995).

The risk communication that took place in the community led to uncertainty and open ends, and the community members and households were forced to navigate between normality and exceptionality. People's responses to this environmental crisis were guided by their interpretations of the situation and its outcomes, as well as their collective history and feelings of belonging. If the communication of risk vacillates between contradictory messages, the risk signal is distorted and open to interpretation. If the local community does not have enough information, or if there is a disagreement about the nature of the hazard and its effects, groups at the local level may develop their own understanding of the crisis and how to cope with the problem (Fessenden-Raden et al., 1987). In response, the risk may either be amplified or downplayed. An environmental crisis entails far-reaching uncertainty because it is difficult to measure and predict the environmental consequences on the local community. The international literature on communities which have been affected by toxic pollution of groundwater and drinking water, for example, or by other manmade disasters resulting from technological impacts, points to the high social and cultural complexity of such situations and attests that sociocultural aspects must be taken into consideration when understanding, assessing, and communicating risk (Fitchen, 1989; Levine, 1982; Sjölander-Lindqvist, 2004).

Situations of uncertainty trigger a collective process of meaning-building because people must make sense, to themselves as well to others, of how the local environment has changed (Sjölander-Lindqvist, 2004), or as happened in this case, they may also turn to their past. Instead of amplifying the risks, they entered a process of elaborating the risks. The local community "saw" the contamination and its risks, but their attachment to the landscape, the place where they live, and their collective history and feeling of belonging also influenced the risk-assessment process and the contradictory communication of risk.

By Way of Conclusion

This study has identified narratives cherishing the mining industry's past contribution to a vital community. The study concludes that by identifying and analyzing how people conceptualize and construe risk in the post-treatment of contaminated places, we learn more about the symbolic meaning of place. As the study has demonstrated, the plans for remediation resulted in the negotiation of the risks associated with living in toxic environments. The members of the local community began negotiating the meaning of place, community, and the role the copper industry had played in the past, but also the role of the past for contemporary community lives, while simultaneously negotiating, and sometimes suppressing, both explicitly and implicitly, the risks associated with arsenic. This understanding proceeded from the formulation and distribution of the vacillating risk messages. Since the place of the local community is both a place for dwelling and a reminder and the result of the lives of past generations, Riddarhyttan becomes a collective experience and a memory based on relationships with a geographical space bestowed with meanings and values. For risk communication to be adequate, the shared information can obviously be neither contradictory nor insufficient in terms of scientific knowledge, nor can it disregard the social or experiential conditions or contexts of communication. With that in mind, the results and conclusions of this case study are a stark reminder of risk as situated in the lives of both the sender and the receiver of the communication (Boholm, 2003).

What the Riddarhyttan case study clearly implicates is that the risk communication process and the willingness to listen and trust the communicator are clearly influenced by symbolic cultural notions and values pertaining to the institutions of home, family, homeownership, community, and history (Sjölander-Lindqvist, 2004). These values and notions emphasize the aspect of time and temporality, since the totality of lived practices and experiences of community members shape the concomitant reactions of communities (Boholm, 2015; Fitchen, 1989; Sjölander-Lindqvist, 2004). This perspective advocates the role of the past in constructing notions of current and potential future lives, in which community residents' cultural values and the meanings associated with the local environment are at the heart of their concerns and understandings. Whereas the risk signal from the sender was distorted by vacillating risk messages, the testimony of the lives and works of earlier and present generations (Inglis, 1977; Ingold, 1993) also informed the communication process. In terms relevant to this study, the concerns of the local community, the locally defined understandings of the copper mining fields, and the ensuing negotiation of normality and exceptionality remind us of the link between experience and existence. It is also a stark reminder that risk communication is subject to symbolic cultural notions and values pertaining to the institutions of home, family, homeownership, and community, and how heritage contributes to current and potential future notions of life (Boholm, 2015; Fitchen, 1989; Sjölander-Lindqvist, 2004).

Acknowledgments I wish to thank the authors in this book and the anonymous reviewers for their constructive feedback and helpful comments on earlier versions of the manuscript. The research on which this book chapter is based was funded by Formas—the Swedish Research Agency for Sustainable Development (Contract No. 2014-8566-28232-16). I would also like to thank Madeleine Prutzer, who assisted with interviews and transcriptions of the interviews.

References

Adger, W., Barnett, J., Brown, K., Marshall, N., & O'Brien, K. (2013). Cultural dimensions of climate change impacts and adaptation. *Nature Climate Change, 3*, 112–117. https://doi.org/10.1038/nclimate1666

Appadurai, A. (1995). The production of locality. In R. Fardon (Ed.), *Counterworks* (pp. 204–225). Routledge.

Argenti, P. (2002). Crisis communication: Lessons from 9/11. *Harvard Business Review, 80*, 103–109.

Bindler, R., & Rydberg, J. (2016). Revisiting key sedimentary archives yields evidence of a rapid onset of mining in the mid-13th century at the Great Copper Mountain, Falun, Sweden. *Archaeometry, 58*, 642–658. https://doi.org/10.1111/arcm.12192

Boholm, Å. (2003). Situated risk: An introduction. *Ethnos, 68*(2), 157–158. https://doi.org/10.1080/0014184032000097713

Boholm, Å. (2015). *Anthropology and risk.* Routledge.

Boholm, Å., & Corvellec, H. (2011). A relational theory of risk. *Journal of Risk Research, 14*(2), 175–190. https://doi.org/10.1080/13669877.2010.515313

Casey, E. S. (1996). How to get from space to place in a fairly short stretch of time: Phenomenological prolegomena. In S. Feld & K. H. Basso (Eds.), *Senses of place* (pp. 13–52). School of American Research Press.

Cutter, S. L., Barnes, L., Berry, M., Burton, C., Evans, E., Tate, E., & Webb, J. (2008). A place-based model for understanding community resilience to natural disasters. *Global Environmental Change, 18*(4), 598–606. https://doi.org/10.1016/j.gloenvcha.2008.07.013

de França Doria, M., Pidgeon, N., & Hunter, P. R. (2009). Perceptions of drinking water quality and risk and its effect on behaviour: A cross-national study. *Science of the Total Environment, 407*(21), 5455–5464. https://doi.org/10.1016/j.scitotenv.2009.06.031

Depoe, S. (2007). Environmental communication as nexus. *Environmental Communication, 1*(1), 1–4. https://doi.org/10.1080/17524030701395871

Edelstein, M. R. (1988). *Contaminated communities: The social and psychological impacts of residential toxic exposure*. Westview Press.

Edelstein, M. R., & Wandersman, A. (1987). Community dynamics in coping with toxic contaminant. In I. Altman & A. Wandersman (Eds.), *Neighborhood and community environments* (Vol. 10, pp. 69–122). Plenum Press.

Fessenden-Raden, J., Fitchen, J. M., & Heath, J. S. (1987). Providing risk information in communities: Factors influencing what is heard and accepted. *Science, Technology & Human Values, 12*, 94–101.

Fischoff, B. (1995). Risk perception and communication unplugged: Twenty years of process. *Risk Analysis, 15*(2), 137–145.

Fitchen, J. M. (1989). When toxic chemicals pollute residential environments: The cultural meanings of home and homeownership. *Human Organization, 48*, 313–324. https://doi.org/10.17730/humo.48.4.j5306054tk7w6813

Geertz, C. (1973). *The interpretation of cultures*. Basic Books.

Gibbs, L. M. (1998). *Love canal*. New Society Publishers.

Halbwachs, M. (1992). *On collective memory* (L. A. Coser, Trans.). Chicago: University of Chicago Press. (Original work published 1952).

Högberg, R. (2015, 17 June). Arsenikavfall en fara i gruvområde. *Fagersta Posten*, p. NA.

Inglis, F. (1977). Nation and community: A landscape and its morality. *The Social Review*, 489–514. https://doi.org/10.1111/j.1467-954X.1977.tb00301.x

Ingold, T. (1993). The temporality of the landscape. *World Archaeology, 25*, 152–174.

Karlsson, J., Segerström, U., Berg, A., Mattielli, N., & Bindler, R. (2015). Tracing modern environmental conditions to their roots in early mining, metallurgy, and settlement in Gladhammar, Southeast Sweden: Vegetation and pollution history outside the traditional Bergslagen mining region. *The Holocene, 25*(6), 944–955.

Kasperson, R. E., Renn, O., Slovic, P., Brown, H. S., Emel, J., Goble, R., Kasperson, J. X., & Ratick, S. (1988). The social amplification of risk: A conceptual framework. *Risk Analysis, 8*, 177–187. https://doi.org/10.1111/j.1539-6924.1988.tb01168.x

Levine, A. G. (1982). *Love canal: Science, politics and people*. Lexington Books.

Lidskog, R. (2011). Scientised citizens and democratized science: Re-assessing the expert-lay divide. In Å. Boholm (Ed.), *New perspectives on risk communication* (pp. 69–86). Routledge.

Mairal Buil, G. (2005). Los conflictos del agua en España. *Nómadas, 22*, 126–139.
Olsson, L. (2014, 5 Nov). Förorenad mark i Kolbäck undersöks. *Västeråsposten*, p. 23.
Pidgeon, N. F., Kasperson, R. E., & Slovic, P. (Eds.). (2003). *The social amplification of risk*. Cambridge University Press.
Power, M. (2007). *Organized uncertainty: Designing a world of risk management*. Oxford University Press.
Rappaport, R. A. (1996). Risk and the human environment. *Annals of the American Academy of Political and Social Science, 545*, 64–74. https://doi.org/10.1177/0002716296545001008
Renn, O. (1991). Risk communication and the social amplification of risk. In R. E. Kasperson & P. J. M. Stallen (Eds.), *Communicating risks to the public: Technology, risk, and society* (pp. 287–324). Springer.
Renn, O., & Levine, D. (1991). Credibility and trust in risk communication. In R. E. Kasperson & P. J. M. Stallen (Eds.), *Communicating risks to the public: Technology, risk, and society* (pp. 175–218). Springer.
Schama, S. (1995). *Landscape and memory*. Alfred A. Knopf.
Shore, C., Wright, S., & Però, D. (Eds.). (2011). *Policy worlds: Anthropology and analysis of contemporary power*. Berghahn.
Sjölander-Lindqvist, A. (2004). *Local environment at stake: The Hallandsås railway tunnel in a cultural and social context*. Doctoral thesis, Lund: Human Ecology Division.
Steelman, T. A., McCaffrey, S. M., Velez, A. K., & Briefel, J. A. (2015). What information do people use, trust, and find useful during a disaster? Evidence from five large wildfires. *Natural Hazards, 76*, 615–634. https://doi.org/10.1007/s11069-014-1512-x
Stoffle, R. W., & Arnold, R. (2003). Confronting the Angry Rock: American Indians' situated risks from radioactivity. *Ethnos, 68*, 230–248. https://doi.org/10.1080/0014184032000097768
Stoffle, R. W., Stoffle, B. R., & Sjölander-Lindqvist, A. (2013). Contested time horizons. In A. Bond, A. Morrison-Saunders, & R. Howitt (Eds.), *Sustainability assessment: Pluralism, practice and progress* (pp. 51–67). Routledge.
Sveriges Radio Västmanland (2015). *Ingen hälsorisk att bo nära förorenat område*. Last retrieved: 10 Dec, 2021. https://sverigesradio.se/artikel/6057647
Tilley, C. (1994). Space, place, landscape and perception: Phenomenological perspectives. In C. Tilley, Y. Christopher, B. Kapferer, & J. Gledhill (Eds.), *A phenomenology of landscape place, paths and monuments* (pp. 7–34). Berg.
van Liedekerke, M., Prokop, G., Rabl-Berger, Kibblewhite, M., & Louwagie, G. (2014). Progress in the management of contaminated sites in Europe. Reference Report by the Joint Research Centre of the European Commission. Luxembourg: Publications Office of the European Union.

Venables, D., Pidgeon, N. F., Parkhill, K. A., Henwood, K. L., & Simmons, P. (2012). Living with nuclear power: Sense of place, proximity, and risk perceptions in local host communities. *Journal of Environmental Psychology, 32*(4), 371–383. https://doi.org/10.1016/j.jenvp.2012.06.003

Zabbey, N., Sam, K., & Onyebuchi, A. T. (2017). Remediation of contaminated lands in the Niger Delta, Nigeria: Prospects and challenges. *Science of the Total Environment, 586*, 952–965. https://doi.org/10.1016/j.scitotenv.2017.02.075

Zimmerman, R. (1987). A process framework for risk communication. *Science, Technology and Human Values, 12*(3), 131–137.

Cultural Transmission in Slovak Mountain Regions: Local Knowledge as Symbolic Argumentation

Ivan Murin

Introduction

Many traditional cultures have found their way to managing their surrounding environment. Over many generations, traditional cultures and local communities have learned that having fish in the rivers and wildlife in the forests depends on not cutting down all the trees and not wiping out the fish stock, but keeping it at a viable level to maintain ecological balance. Through their ongoing interaction with the environment and the ways they talk about and with their lands and the species within them, traditional cultures have developed cultural strategies for resource management (see Kugo, chapter "Community Voices, Practices, and Memories in Environmental Communication: Iliamna Lake Yup'ik Place Names,

I. Murin (✉)
Department of Social Studies and Ethnology, Matej Bel University, Banská Bystrica, Slovakia
e-mail: ivan.murin@umb.sk

A. Sjölander-Lindqvist et al. (eds.), *Anthropological Perspectives on Environmental Communication*, Palgrave Studies in Anthropology of Sustainability, https://doi.org/10.1007/978-3-030-78040-1_4

Alaska" of this book). Such culturally defined knowledge includes and builds upon a diverse set of practices, discourses, and ideas, which are shared across generations. This knowledge provides an important foundation for cultural identity at local, regional, and even national levels, but is also a kind of knowledge that is important for sustainability.

Traditional systems of land use, for example, are an important foundation for community viability (Berkes & Folke, 2002), and should therefore be harnessed, as this knowledge can benefit the management of common property. Maintaining Traditional Ecological Knowledge (TEK) is, however, dependent on both the ability of local communities to uphold practices and learn from their performance, and the verbal and intergenerational sharing of stories and narratives about local life, community resources, knowledge, and experiences (Anderson, 2015).

This chapter brings us to the remote Mountainside Settlements of Central Slovakia (MSCS), where the change and abandonment of traditional land-use practices, due to outmigration to urban centers and aging rural populations, has implications for the ability of the *Vrchár*[1] culture to sustain social ties, manage community-based agricultural resources, and protect the cultural landscape. Outmigration has led to depopulation and consequently fewer people who farm the lands of the MSCS. Subsequently, the abandonment of traditional ways of managing land and local resources has led to fewer opportunities to share knowledge and establish a common reference to the environment, which we know come into being through concerted, fortuitous, embodied, and lived practices (Sjölander-Lindqvist & Sandström, 2019).

Depopulation of less productive, remote, and mountainous regions and areas is a trend well known to many countries around the world. In Europe and beyond, recent decades have seen a general demographic phenomenon of the depopulation of peripheral areas, especially those characterized by rural traditions (MacDonald et al., 2000), and migration to industrial centers (Feranec et al., 2010). In such a context, the loss of Traditional Ecological Knowledge may be an obvious outcome. Some of the negative consequences relate to the reduced diversity of the land-use mosaic and consequent landscape homogeneity (Navarro & Pereira, 2012). With the gradual disappearance of local cultures, society loses diversity and cases of adaptation to local circumstances that we can learn from for the

[1] Slovak, sing. *vrchár*, pl. *vrchári*, lit. "mountainers," loosely equivalent to "mountainfolk" or even "hillbillies," given the term's historically pejorative connotation.

development of sustainable solutions. Owing to the relationships between different environments, the constantly evolving changes to the ecosystem, and the ability of a local community to manage local resources, the generational exchange of experiences, insights, and knowledge presents a challenge to contemporary society.

Traditional Ecological Knowledge as an Adaptation Process

Traditional Ecological Knowledge (TEK) is experience that has been acquired from direct human interaction with the environment. It can be thought of as the result of people in natural environments who, through their engagement with the environment and the species within it, have found ways to maintain their livelihoods and thereby discover and develop paths of adaptation (Berkes, 1993; Colloff et al., 2020). In rural areas, TEK is particularly important for "the regulation and balance of exploitative pressures that permit an ecosystem to maintain stability and regenerative capacity" (Ruddle, 1993, p. 17). There are a number of examples in the anthropological literature which show that knowledge is situated and culture-specific. Particularly relevant to the present case study are examples of the development of mountain grazing in similar mountainous conditions in Europe, namely in the Pyrenees (Fernández-Giménez & Fillat Est, 2012), Alps (Gilck & Poschlod, 2019), and Carpathians (Novák et al., 2014). Gradually, geographically distinctive grazing and farming techniques have developed and adapted in various ways to ensure the sustainability of ecosystems and human settlements.

TEK is experience and knowledge, but it is also an intellectual activity found in a wide range of social, cultural, and environmental contexts (Folke, 2004; Gómez-Baggethun & Reyes-García, 2019). The persistence of this knowledge is dependent on it being shared across generations (Cocks, 2006; Ruddle, 1993) according to certain structural and processual principles (Cocks, 2006; Ruddle, 1993) and based on the constant and rigorous perception of changes in society and environment (Colloff et al., 2020). The success of these socio-cultural processes can be assessed in the extent to which individual communities and cultures have managed to co-adapt to their environments (Stoffle et al., 2003). A precondition for successful co-adaptation is the ability of the individual community/culture to, continuously and over time, balance and develop their actions in response to biophysical changes and ecological processes as well those changes brought on by societal processes (Gardner, 2017), but it also requires a continuous flow of knowledge between community members (Cocks, 2006; Ruddle, 1993). During the transmission of knowledge, customary ways of doing things are a core institution through which traditional ecological and environmental knowledge is given and received in a social world and in analogy with "the biological-physical world with which it overlaps" (Ruddle, 1993, p. 19). This knowledge is, to borrow from Ruddle (1993), a "key dialectic of society," as traditionally and culturally grounded knowledge is about being—socially, culturally, and ecologically—in the world. From this perspective, TEK is a concept that captures both the everyday social reality and our continuous reproduction of everyday life (Berger & Luckmann, 1966).

Ultimately, the persistence of TEK is dependent on continuous transmission over generations. Turning our eyes to contemporary life, in Slovakia, Europe more broadly, and other parts of the world, we find society struggling with the challenges of globalization, urbanization, digitalization, individualization, and demographic change. These have far-reaching consequences for human well-being, which requires the integration of traditional knowledge to enable the development of sustainable solutions to the key environmental threats of climate change, land degradation, and loss of biodiversity (IPBES, 2019). The long-term trend of aging rural populations is in itself worrying, since it, together with urbanization, increases the rate of depopulation, reflecting in part the tendency of young people to leave the countryside to find employment and another kind of life in urban and metropolitan areas throughout Europe (The Permanent European Conference for the Study of the Rural Landscape). In addition, rural areas in Europe are experiencing a growing share of young unemployed people (aged 18–24) who do not continue their education past the secondary level (Eurostat, 2019). These developments have consequences for the capacity of the rural communities to adapt to changing circumstances (Lieskovský et al., 2015). While it has long been assumed that the level of economic development is crucial to a society's capacity to adapt to climate change, evidence points toward the conclusion that societies that depend "more on experience, knowledge and [...] weather-sensitive resources" are better at finding solutions to environmental risks (Adger & Vincent, 2005).

Anthropological theory on how people and cultures adapt to different circumstances (Steward, 1955; McDonnell & Pickett, 1993; Ruddle,

1993; Stoffle et al., 2013) is important with respect to TEK transmission, together with anthropological knowledge on how we, as human beings, create a sense of belonging to particular places and landscapes as bounded spaces for thought and action (Appadurai, 1986; Ingold, 2000; Sjölander-Lindqvist & Sandström, 2019; Kugo, chapter "Community Voices, Practices, and Memories in Environmental Communication: Iliamna Lake Yup'ik Place Names, Alaska" of this book; Sjölander-Lindqvist, chapter "Arsenic Fields: Community Understandings of Risk, Place, and Landscape"; Stoffle, chapter "Living Stone Bridges: Epistemological Divides in Heritage Environmental Communication"; Van Vlack, chapter "Dancing with Lava: Indigenous Interactions with an Active Volcano in Arizona"). These two areas of theory bring important perspectives to the role of TEK in co-adaptation, but how we transmit information through symbolic references to the environment is also important in terms of environmental communication. Through knowledge of partial adaptations, the competency for global sustainability policy increases. From an anthropological standpoint, cultural adaptations are considered more important than technological adaptations. This can be justified by the fact that the creation of symbolic references to the environment, classifications, and taxonomy is a long-term process, a consequence of the daily contact of generations of people with the environment. The cultural ways of co-adaptation are the basis for the moral order of societies; they are explained in narratives and transmitted in symbols.

Methods

In 2005, local action groups contacted Vrchári from the northern Malohont region. These people represented a generation of Vrchári who, at the end of the twentieth century, had the lands of their ancestors returned to them. They came from different regions of Slovakia and did not have direct-contact experience with the environment that their ancestors had inhabited. They were interested in revitalizing their ancestors' environment, but lacked the necessary knowledge of the practices in which their ancestors had worked in this environment.

People in the local communities invited anthropologists to (1) conduct archival research into Vrchár culture, (2) compile an inventory of the current state of Vrchár culture, and (3) incorporate the local TEK into applications for environmental sustainability grants. The archival research provided sufficient historical findings of TEK workers in northern

Malohont. In 2010–2017, the anthropologists selected the Northern Malohont Research Area as a site for research into the continuity and discontinuity of cultural transmission. The real, manifest, as well as latent contents of the lived culture's TEK were investigated by inventorying (Horton et al., 2017) the area. Via observations and interviews, the predominant modes of cultural transmission were identified (LeCompte & Schensul, 2013). With respect to the revitalization possibilities, a fourth action was also taken: creating a TEK transmission model (Table 1) for the new generation of Vrchári with a view toward sustainability for subsequent generations.

The grant application component focused on developing a local theory of the current population of Vrchári and on their new strategies of communication with the public. The applicants and partners are

Table 1 Generalized model of TEK (Babai & Molnár, 2014, p. 126)—Traditional treatments of hay meadows (A1–A8) and pastures (B1–B4) by older generations of Vrchári in northern Malohont, Slovakia

Meadow/pasture	*Feb*	*Mar*	*Apr*	*May*	*Jun*	*Jul*	*Aug*	*Sept*	*Oct*	*Nov*	*Affected area (% of total)*
A1a—mowing				x		x		(x)			>40
A1b—mowing and regrowing[a]				x							>50
A2—clearing		x									>60
A3—hayseed scattering					x						10–15
A4—oversowing with *Onobrychis*		x									0.1–0.5
A5—extensive manuring	x										25–30
A6—weeding								x			30–40
A7—moss suppression	x				x						0.1–0.5
A8—marsh draining			x								0.1–0.3
B1—grazing				(x)	x	x	x	x	x	(x)	>95
B2—pasture clearing	x	x			(x)	x					70–75
B3—fence repair	x						x		x		>95
B4—burning to prevent *Nardus*	x	(x)									5–10

[a]*In autumn cattle graze on aftermath*

multi-stakeholders, mainly government representatives and advocates of intensive forest management and intensive agricultural development. Symbolic arguments (Depoe, 2007; Cox, 2006, 2015) of continuity and discontinuity of the MSCS are gradually gaining general support.

Mountains and Vrchári

Historical geographers consider the MSCS to be the best preserved type of historical structure of traditionally farmed agricultural landscapes in Slovakia. In ethnography, the displaced and dispersed forms of MSCS are called *lazy*, *vrchy*, *kopanice*, *štále*, and *rale*. Their main identifying characteristic is dispersed (family) settlements (DSs). The culture that has emerged in this way is called Vrchár culture (*vrchárska kultúra*). In Slovak, its bearers are called *vrchári*, *lazníci*, or *kopaničiari*. DSs are an autonomous demonstration of Vrchár culture and socioeconomic activity, preconditioned by specific natural and historical conditions. In the text, we will call them by their Slovak name, Vrchári, as used in ethnological studies (Ethnographic Institute of the Slovak Academy of Sciences, 1994). Anthropologically they are considered an ethnographic group with their own mountain-adapted culture. The map in Fig. 1 displays the Vrchár culture in the Slovak Ore Mountains (*Slovenské rudohorie*) and Krupina Plain (*Krupinská planina*) as of 2018 (92 municipalities over an area of 2258 km^2).

The emergence of Vrchár culture is related to several attempts to settle the mountainous areas of Central Slovakia from the twelfth to the nineteenth centuries. Gradually, more multigenerational homesteads were established on the outer edges of the valley villages. The emergence of almost continually inhabited landscape areas was related to the internal migration within the mountain and foothill areas of Slovakia which took place in the sixteenth–nineteenth centuries (Huba, 1997). The Vrchári adapted to the mountain environment via mountain farming and other complementary socioeconomic activities. The Vrchári gradually developed a specific type of land exploitation, a type of settlement, and an intrinsically identifiable culture. Their origin is characterized by (1) efforts to use the land of the forest and ecumene more efficiently on the one hand, and on the other, by (2) the stimulating effect of human adaptation on population growth. Settlement waves were the directly connected to the relative overpopulation of Slovakia's historical ecumene. The establishment of DSs also had a number of sub-causes, such as higher soil cultivability in the case

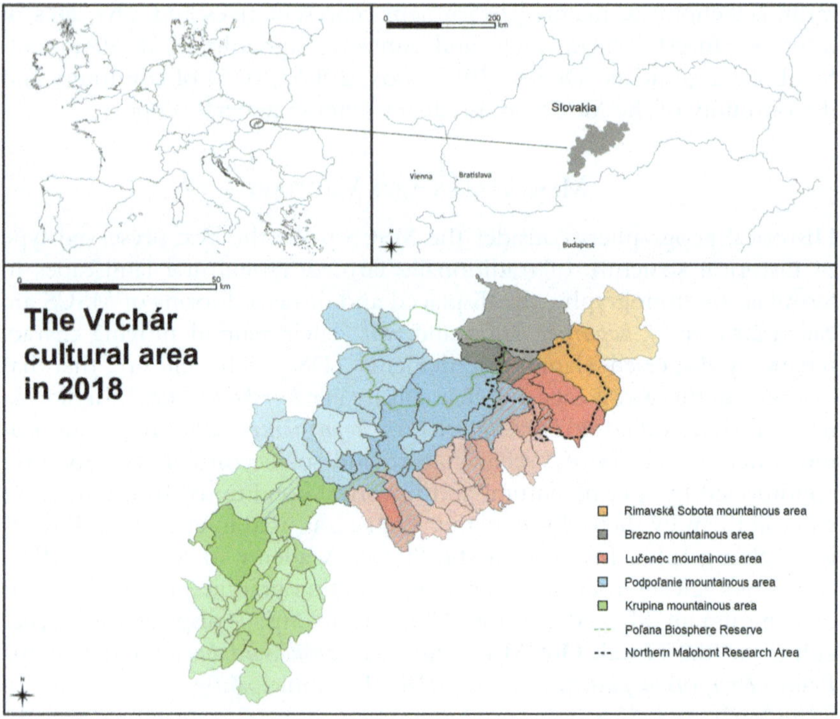

Fig. 1 The Vrchár cultural area in 2018, shaded according to the occurrence of significant cultural features (Ethnographic Institute of the Slovak Academy of Sciences, 1994). The main areas of Vrchár culture (darkest areas) originated outside dense settlements from the seventeenth to the nineteenth centuries. Via young internal migration in the nineteenth century, the DSs also spread to adjacent areas (medium-dark areas). The lightest areas are those DSs that originated in the twentieth century and later lost the original function of multigenerational homesteads. Cultural overlaps of the sub-regions are marked by hatching. (This map was created by Michal Filadelfi and Ivan Murin in QGIS (free version https://qgis.org/en/site/) and Adobe, license no. CE08041211, and is used with its creators' permission)

of permanent settlement of remote parts of the villages, relative abundance of free land, lower taxes, search for refuge from war, and so on (Petrovič, 2006).

A multi-member three-generation family became the basis for the sustainability of the emerging mountainside settlements. Its numerosity directly depended on the ability to process the soil. However, the relationships between the settlements were most affected by the crucial reciprocity between the inhabitants. This was characterized by extensive forms of interchange between families linked by closer ties. This led to internal integration within Vrchár society, resulting in a culture that was cohesive and set apart from urban society. Social norms and cultural values were created independently of the populations of the agrarian lowlands and towns. There was a prevalence of direct-contact generational transfer of knowledge and its synchronization with changes in the environment (Murin & Kandert, 2018).

The greatest expansion of Vrchár culture can be identified in the first half of the twentieth century,[2] triggered by the First Land Reform of Slovakia at the beginning of the twentieth century. A large land property was allocated to the Vrchári, by which the area occupied by DSs expanded (Špulerová et al., 2016). During this period, the MSCS had high structural diversity and therefore ecological stability. Due to their heterogeneity of shape and area, and varied cultivation of culturally characteristic crop varieties, the model of DSs was created to describe the MSCS. It was during this period that the cultural character of the Vrchári as an ethnographic group took shape.

The development in the Vrchár settlement area in the second half of the twentieth century was mainly associated with extensive industrialization. This constituted the first violation of the Vrchár generational structure (Petrovič, 2006). The increasing individualization of activities and, in particular, the compulsory employment of Vrchári in state-owned or large-scale agribusinesses and the metals industry was the reason for the first significant departure of the young generation. This change was adopted

[2] The first mentions of the number of villages with dispersed settlements come from Janšák (1929), who identified 132 villages with 2176 parts. In 1961, Slovakia was home to 166 villages with dispersed settlements with 2899 dispersed parts on an area of 4640 km^2, which made up 9.46% of Slovakia. The increase in the number of villages and dispersed settlements is probably the result of inaccurate census-taking in 1929 and not the establishment of new villages. At present, the emergence of new villages with dispersed settlements is unlikely, even unimaginable, given the lack of new farmland. Originally, dispersed settlements displayed a prevailing tendency of decline and extinction. The basic function of the dispersed settlements was soil cultivation in marginal areas. In recent years these areas have been in decline (Mládek et al., 2006).

by the Vrchár community, giving rise to a new model of the Vrchár as "iron-farmer" (*kovoroľník*) (Kandert, 2007).

Gradually, however, state pressure for Vrchári to leave the MSCS grew stronger. The totalitarian ideology of Communist Czechoslovakia justified the decline of DSs as industrial and civilizational progress. Two other state/political decisions in the second half of the twentieth century significantly marked the development of DSs and the culture of the Vrchári. The strategy of intensifying agriculture and forestry led to the consolidation of the disconnected, scattered homesteads. This also extended to terrain, water management, and other measures to increase soil fertility and improve economic returns.[3] State-managed collectivization and cooperative farming led to the extinction of many DSs and the overall homogenization of the countryside (Lieskovský et al., 2015). Following voluntary resettlement in the 1950s and the subsequent state-initiated displacement of Vrchári in the 1970s, a large part of the MSCS was left abandoned. At the beginning of the 1980s, Czechoslovak ethnologist Soňa Švecová expressed the opinion that the DSs would either go extinct as a consequence of Slovakia's new socioeconomic situation or exist in other (recreational) conditions and functions (Švecová, 1975).

After 1990, the displacement of the Vrchári was partially halted and the landscape structure of MSCS became more diversified. There was some generational rejuvenation of Vrchári and loss of landscape macrostructures. This was caused by a change in the land ownership structure when the land was returned to its original owners or their heirs.[4]

Land Abandonment as Loss of Cultural and Natural Diversity

At present, DSs rarely fulfill their primary function of providing permanent residence to multigenerational families. The communities of the Vrchári are not territorially concentrated, and their social ties have been lost. Their main focus is on cottage industry and productive farming

[3] Land modification during the period of socialization of Slovak agriculture was carried out in two stages. The first stage was implemented in 1955 via Regulation No. 47/1955 Coll. The second stage of land consolidation was guided by Regulation No. 12 064/1977 in 1977.

[4] Act No. 229/1991 Coll. on the Modification of Ownership Relations to Land and Other Agricultural Property abolished all Communist-era land expropriations and restored full ownership to the original owners.

(Priečko, 2015). Cultivation is mostly limited to maintaining permanent grassland near the houses, especially for aesthetic purposes, or fruit orchards. More sustainable are those areas which have retained a diversity of resources and income. This is mainly due to residues from mountain farming and the provision of ecosystem services (Sandhu & Wratten, 2013). However, the young generations of Vrchári lack the continuous TEK they would need in order to be successful.

A large part of the land is only sporadically used, exhibiting ecological succession and overgrown by woods and shrubs (Kanianska, Kizeková, Nováček, & Zeman, 2014). The original expectations of mid-twentieth-century central forest planners that the intensification of forest cultivation would create new livelihood opportunities for Vrchári have not been fulfilled. Planting forest for cultivation can only feed a few workers skilled in forestry. Wood is not processed locally, but transported to industrial centers.

At present, the significant risk to the Vrchári's environment and population is global warming. It is causing rapid extinction of the uninhabited forest plantations and abandonment of the countryside by the last Vrchári (Lieskovský et al., 2015), as you can see in Figs. 2 and 3. In the foothills, it is mainly farmed land and permanent grassland that are being lost.[5] In this way the epicenters of species diversity (including the human population) and cultural diversity are disappearing.[6] The Vrchári's mowed meadows and extensive pastures are among the richest plant communities in this area. However, their maintenance is not possible without regular care. A major problem is the expansion of high grasses, which push out other herbaceous vegetation. This vegetation is the most valuable trait of herbaceous grassland communities, and it is also lost to colonization by woods.

In 1992, the Slovak Republic returned the land in the northern Malohont region to the previous Vrchár landowners from before the Communist era. For anthropologists, this created an opportunity to explore the re-adaptation of new generations to the environment of their ancestors. Such research has become the basis of cooperation between

[5] The migratory front of the ravines grows at a rate of about 1.7 m per year. Since the countryside was abandoned and mowing was discontinued, the edge of the woodlands has increased by more than 80 m in about 50 years.

[6] According to a comparison of historical aerial photographs over the last decade, we know that the rate of change is not constant, but accelerating. Clonal species grow in the herbaceous layer, and the grasslands are ceasing to be a part of the landscape image of the DSs.

Figs. 2 and 3 Lived-in (Fig. 2, left) and abandoned (Fig. 3, right) DSs in the 1970s and 1980s, and planted monoculture of common spruce (*Picea abies*). (These images used with permission of the Slovak Academy of Sciences, Institute of Ethnology and Social Anthropology, 1977, No. AD008941)

researchers and Vrchári in the application of TEK arguments to environmental communication.

Anthropological Arguments for the Continuity of Generational Transmission

When explaining the patterns of culture transfer, we often point to parallels with the equivalents of general population evolution. With human populations, we are looking for a common culture origin, selection, and adaptation processes for the constant change, stabilization, and subsequent variation of culture in its various reproductions. Transmission of culture can be considered to be the distribution of culturally specific information bound to the local community (Fig. 4) which is transmitted through various transmission modes among its members (Tindall, 1976; Cavalli-Sforza & Feldman, 1981). We take this to mean that this information has a limited scope and links to local contexts. It originates in well-known sources, follows a consistent dissemination algorithm, and is applied in similar community structures and in similar historical contexts. Transmission in a scattered environment is perceived thus: the more closely people's lives are intertwined and the longer they share similar

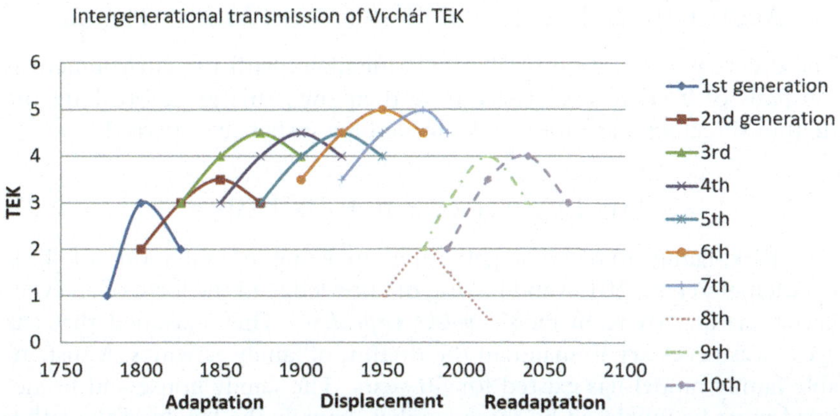

Fig. 4 The six-function adaptation scheme (Stoffle et al., 2003) and discontinuity of Vrchár culture. (Created by Ivan Murin)

contexts of environment and historicity, the higher the number of common cultural elements they display.

Argument 1: The Floating TEK Gap

Northern Malohont was forested 50 years ago, and nowadays it is covered by spruce forest plantations. Due to global warming, it is endangered by wind, drought, and ligniperdous insects *(Ips typographus)*. In the past, working in the woods was one of the seasonal jobs of the Vrchári. At present it is their only work. The most common images of DSs are family homes transformed into garages of forest mechanization. The transformation of the last Vrchári into loggers was generally caused by a massive rise of natural disasters in large forest plantations (Gömöry et al., 2006). Consequently, young generations are not abundantly integrated in alternative survival activities in DSs.

> When I left for the city, I knew how to do everything I touched. What I didn't know, I learned quickly. For life in the mountains, you needed such people. The young ones in the mountains now are helpless. They can only do one job, they didn't have the opportunity to learn everything, so they're poor. (Vrchár male, 60 years, interview, 2018, trans. Ivan Murin)

Argument 2: The Three-Generation Model Family

The everyday interaction of all family members with the environment is the basis of TEK in use. These have their own rhythm. Disturbing the rhythm threatens the entire circle needed for sustainable survival.

The Example of the Ilčík Family

The Ilčík family initiated a gathering involving presentations of TEK (Lindenfeld et al., 2012) and linking of knowledge in the form of an event known as an "ox-team race" (*volské záprahy*).[7] They reasoned that the event was necessary to maintain the rhythm of family activities. A sustainable family model has existed for 50 years. The family house and homestead must be inhabited by three families, namely the family of the father and the families of his two children. In addition to working in the field, they must also have jobs that do not require more than eight hours outside the homestead. Single members may leave for work for longer. Men and members of the youngest, still single generation must be able to commute to towns. Married women and some single young women may be seasonally employed outside the DS.

> To stay here in the hills, we have to be a big family. Every hand and every mind is missed when it's gone. We all have to know what we have to do every day. Not everyone gets used to it and learns. When the old ones die, we won't know how to go on. When the young ones leave, there won't be anyone left to work. (Vrchár female, 82 years, interview, 2018, trans. Ivan Murin)

Any disturbance of the family rhythm threatens Vrchár culture. Each loss of knowledge must be replaced by something from the external environment. Each time an activity is isolated from its context, there is a loss of the complexity of TEK.

[7] Originally denoting an actual ox-team race, *volské záprahy* now refers to a festive event involving, among other activities, a stylized procession of a team of oxen. This newfound tradition has led to a modest revival of ox husbandry and the oxen's status among Vrchári, with some even keeping them as work animals once again.

Argument 3: Cattle Farming as a Symbol of Identity, Integrity, and Livelihood

The DSs provide conditions for family cattle breeding. The number of cattle kept during the winter is directly related to the sufficiency of the stocks of fodder, especially hay. Mowing, drying, and storing hay is one of the most important activities in the farmer's economic year, and the whole family takes part. Higher-quality meadows are usually mowed twice a year. An important piece of TEK is the maintenance of meadows: clearing them, collecting stones in the spring and after mowing, and fertilization. The process also includes transporting the dried hay, and the TEK of conveying the hay from the steep slopes to the DSs with a team of oxen has been preserved.

Grazing, milk and meat production, and the oxen yoke became the symbols of the world of the Vrchári, representing their source of income and raw ingredients for food preparation. Prosperity was associated with seasonal cattle for sale and high cow's milk production. Farming with three to four head of livestock requires intensive relationships with five to six families with the same ways of husbandry. Collaboration then creates synchronic activities that are not substitutable.

> We need to keep a few neighboring families in order to stay alive here. They must have four to five cows, like us, and at least six family members to take care of them. We don't have grandparents who could stay with the cattle while we go to town. Common grazing saves us all time. After some young people left for the towns, we had to abandon the common pasture. The new generation hasn't been able to learn fast and effective ways of building fences in the rugged terrain. (Vrchár male, 40 years, interview, 2018, trans. Ivan Murin)

Each Vrchár cultural community has symbols that represent their identity and mutual integrity. For the Northern Malohont Research Area, it is cattle. The team of oxen is historically the most important means of transport there (Fig. 5). This was reflected in self-identifying activities.

> When we were being moved out of the hills, children and furniture went first. Then we took clothes, quilts, and rugs. My husband and I went last. He had a team of oxen. In the wagon, there was food just in case. I took our cow so we could get through the beginnings. In the city, it was all useless,

Fig. 5 Symbolic procession representing the vanishing cattle-breeding tradition among Vrchári. (This image used with permission of Vojtech Ilčík)

we had to start all over. (Vrchár female, 82 years, interview, 2018, trans. Ivan Murin)

New generations of Vrchári suffer from a reduced tolerance of very high amounts of milk in their diet. Children are seen with lactose intolerance, which was only rarely seen among older generations of Vrchári. In terms of pragmatic arguments, this leads us to consider relatively recent findings about transgenerational epigenetic inheritance—the influence of a person's environment on their behavior, their physiological processes, and those of future generations as well. An oft-cited example of such biocultural evolution is human digestion of lactose (Itan et al., 2010).[8]

[8] The ability to digest lactose past infancy is apparently caused by a single mutation in the regulatory sequence of the dominant gene that governs this process (Lukito et al., 2015). This mutation is more prevalent in areas with a long-established tradition of raising cattle and consuming dairy products (Gerbault et al., 2011). This fact may prematurely lead us to the conclusion that selective pressures governed by culture are much stronger than others, which can be explained by the ability of a cultural trait to, through social learning, spread much

Typically, the ability to digest dairy products is higher in populations with a long cultural history of consuming, processing, and preserving milk. Thus, one pragmatic argument for environmental sustainability is literally rooted in the foothill pastures of northern Malohont. A symbolic argument, on the other hand, is embodied in the animals themselves: cows and oxen as symbols of vitality.

Discussion

One of the main sustainability changes around the world are changes in the generational structure of local populations, the phenomenon of aging, and other demographic indicators. In this case we have outlined secondary social-ecological processes which started and developed in parallel with these changes in peripheral areas in Slovakia (Mládek et al., 2006). Through anthropological research, we can recognize ***what impact this phenomenon has on long-established communities and its implications for sustainability*** (Depoe, 2007; Cox, 2006). With the disappearance of many local cultures, or in better cases, their preservation by museums, we have lost examples of diversity and bases for sustainability. Examples of adaptations have a cultural value when we can verify their sustainability features in the living system. Logic, thinking, and decision making in everyday interactions with the environment form an alternative epistemological basis.

Recent discussions on the argumentation used in environmental communication (Anderson, 2015; Besely, 2015; Cox, 2015) further support the efforts of anthropologists to broadcast "native voices" from their fieldwork. Voices from field research are met with understanding wherever change-oriented research is problem-driven and targets processes of social change. However, dilemmas still persist as to how to seriously reflect on native knowledge in the broad causality of nature and culture. The dilemma presented by a joint team of researchers from Uppsala and Exeter (Joosse et al., 2020) is the issue of the acceptance of co-producing (sustainability) knowledge. They point out (p. 7) that, in their study, "hierarchical and bureaucratic logic pushed most ambitions and ideas for future development aside," which is also observable in the present study and

faster than a genetic mutation; as more and more individuals adopt the trait, the genes that encourage it are subjected to ever more intensive pressure (Švorcová & Kleisner, 2017, p. 353).

beyond. On the other hand, we have encouraging findings on the acceptance of native knowledge where policymakers are increasingly acknowledging the importance of native knowledge (indigenous knowledge, TEK) for understanding and adapting to environmental change (Armitage et al., 2011; Raymond-Yakoubian & Daniel, 2018; Wheeler et al., 2020).

In interviews with members of local communities, we discovered local voices illustrating ***what communication arguments are used in activities to maintain their culture***. Members of revitalized Vrchár communities are aware of the need to foster cross-generational cultural transfer. They create new incentives to preserve it and have set up a series of activities aimed at local stakeholders and their own community members, as well as visitors to revitalized areas. The conservation of generative transmission processes is considered to be a method of preventing ***generational amnesia*** of TEK (Cantrill, 2010, p. 29). Seasonal events have increased public interest. Vrchári have been able to expand regional education to include teaching schoolchildren about Vrchár culture, and they run regular workshops for the potential new generation of Vrchári.

"***What is the cultural, symbolic, and pragmatic value of the increased activism*** of the last generations of depopulating European peripheries?" is a question often faced in environmental communication. In applying the findings of anthropologists, environmental communication scholars work with the thesis that effective environmental communication constitutes a trigger for implementing aid to stabilize or revitalize culture (Schwarze, 2007). Its success is based on authentic experiences, their distinctive epistemological origins (Stoffle & Minnis, 2008), and symbolic argumentation. Based on the research in Slovakia, we contend that effective environmental communication is not based on the absolute number of argumentative communicators. The value that will always appear in a reflective dialogue is the long-term, deep, everyday experience of community members with their co-adaptation to a specific environment.

Acknowledgments This chapter and the research it is based on were supported by public funding from the Slovak Arts Council as part of the project *The Presentation of Research, Condition and Protection of Traditional Culture During the European Year of Cultural Heritage*, No. 18-460-01673, Slovak Ministry of Culture.

References

Adger, N., & Vincent, K. (2005). Uncertainty in adaptive capacity. *Comptes Rendus Geosciences, 337*(4), 399–410.

Anderson, A. (2015). Reflections on environmental communication and the challenges of a new research agenda. *Environmental Communication, 9*(3), 379–383. https://doi.org/10.1080/17524032.2015.1044063

Appadurai, A. (1986). *The social life of things: Commodities in cultural perspective.* University Press.

Armitage, D., Berkes, F., Dale, A., Kocho-Schellenberg, E., & Patton, E. (2011). Co-management and the co-production of knowledge: Learning to adapt in Canada's Arctic. *Global Environmental Change, 21*(3), 995–1004.

Babai, D., & Molnár, Z. (2014). Small-scale traditional management of highly species-rich grasslands in the Carpathians. *Agriculture Ecosystems & Environment, 182*(1), 123–130.

Berger, P., & Luckmann, T. (1966). *The social construction of reality: A treatise in the sociology of knowledge.* Doubleday.

Berkes, F. (1993). Traditional ecological knowledge in perspective. In J. Inglis (Ed.), *Traditional ecological knowledge: Concepts and cases* (pp. 1–9). Canadian Museum of Nature/International Development Research Centre.

Berkes, F., & Folke, C. (2002). Back to the future: Ecosystem dynamics and local knowledge. In L. Gunderson (Ed.), *Panarchy: Understanding transformations in human and natural systems* (pp. 122–137). Island Press.

Besely, J. (2015). Making environmental communication work: Creating useful guidance. *Environmental Communication, 9*(3), 398–403. https://doi.org/10.1080/17524032.2015.1044006

Cantrill, J. G. (2010). Measurement and meaning in environmental communication studies: A response to Kassing. *Johnsson, Kloeber, and Wentzel, 4*(1), 22–36.

Cavalli-Sforza, L. L., & Feldman, M. W. (1981). Cultural transmission for a continuous trait. In *Cultural transmission and evolution: A quantitative approach* (pp. 346–351). Princeton University Press.

Cocks, M. (2006). Bio-cultural diversity: Moving beyond the realm of indigenous and local people. *Human Ecology, 34*(2), 185–200.

Colloff, M., Wise, R., Palomo, I., Lavorel, S., & Unai, P. (2020). Nature's contribution to adaptation: Insights from examples of the transformation of social-ecological systems. *Ecosystems and People, 16*(1), 137–150.

Cox, R. (2006). *Environmental communication and the public sphere.* SAGE.

Cox, R. (2015). Scale, complexity, and communicative systems. *Environmental Communication, 9*(3), 370–378.

Depoe, S. (2007). Environmental communication as nexus. *Environmental Communication, 1*(1), 1–4.

Ethnographic Institute of the Slovak Academy of Sciences. (1994). *Ethnographic atlas of Slovakia: Translations and explanations of texts.* VEDA.
Eurostat. (2019). https://ec.europa.eu/eurostat
Feranec, J., Jaffrain, G., Soukup, T., & Hazeu, G. (2010). Determining changes and flows in European landscapes 1990–2000 using CORINE land cover data. *Applied Geography, 30*(1), 19–35.
Fernández-Giménez, M., & Fillat Est, F. (2012). Pyrenean pastoralists' ecological knowledge: Documentation and application to natural resource management and adaptation. *Human Ecology, 14*(2), 287–300.
Folke, C. (2004). Traditional knowledge in social-ecological systems. *Ecology and Society, 9*(3), [online]: http://www.ecologyandsociety.org/vol9/iss3/art7/
Gardner, A. (2017). The purpose of adaptation. *Interface Focus, 7*(5), Epub 2017 Aug 18.
Gerbault, P., Liebert, A., Itan, Y., Powell, A., Currat, M., Burger, J., Swallow, D. M., & Thomas, M. (2011). Evolution of lactase persistence: An example of human niche construction. *Philosophical Transactions: Biological Sciences, 366*(1566), 863–877.
Gilck, F., & Poschlod, P. (2019). The origin of alpine farming: A review of archaeological, linguistic and archaeobotanical studies in the Alps. *The Holocene, 29*(9), 1503–1511.
Gómez-Baggethun, E., & Reyes-García, V. (2019). Reinterpreting change in traditional ecological knowledge. *Human Ecology, 41*(4), 643–647.
Gömöry, D., Dovčiak, M., Gömöryová, E., Hrivnák, R., Janišová, M., & Ujházy, K. (2006). *Demecological, synecological and genetic aspects of colonization of non-forest areas with forest trees.* Technická Univerzita vo Zvolene.
Horton, C., Hall, D., Gilbertz, S., & Peterson, T. (2017). Voice as entry to agriculturalists' conservationist identity: A cultural inventory of the Yellowstone River. *Environmental Communication, 11*(5), 609–623.
Huba, M. (1997). Kopaničiarske osídlenie, životné prostredie a trvalo udržateľný spôsob existencie [Dispersed Kopanitse settlement, environment and sustainable way of existence]. *Životné prostredie, 31*(3), 61–66.
Ingold, T. (2000). *The perception of the environment: Essays on livelihood, dwelling and skill.* Routledge.
IPBES. (2019, December 20). *Global assessment report on biodiversity and ecosystem services.* https://ipbes.net/global-assessment
Itan, Y., Jones, B. L., Ingram, C. J., Swallow, D. M., & Thomas, M. G. (2010). A worldwide correlation of lactase persistence phenotype and genotypes. *BMC Evolutionary Biology, 10*(1), 1–11.
Janšák, Š. (1929). Príspevok k štúdiu osídlenia Slovenska [A contribution to the study of the settle colonize of Slovakia]. *Sborník muzeálnej slovenskej spoločnosti, 23*, 93–111.

Joosse, S., Powell, N., Bergeå, H., Böhm, S., Calderón, C., Caselunghe, E., Fischer, A., Grubbström, A., Hallgren, L., Holmgren, S., Löf, A., Nordström Källström, H., Raitio, K., Senecah, S., Söderlund, C., Von Essen, E., Westberg, L., & Westin, M. (2020). Critical, engaged and change-oriented scholarship in environmental communication: Six methodological dilemmas to think with. *Environmental Communication, 14*(6), 758–771. https://doi.org/10.1080/17524032.2020.1725588

Kandert, J. (2007). Local history and social networks in everyday political practice: The case of Central Slovakia and Southern Moravia. In K. Roth (Ed.), *Soziale Netzwerke und soziales Vertrauen in den Transformationsländern: ethnologische und soziologische Untersuchungen* (pp. 213–218). Lit Verl.

Kanianska, R., Kizeková, M., Nováček, J., & Zeman, M. (2014). Land-use and land-cover changes in rural areas during different political systems: A case study of Slovakia from 1782 to 2006. *Land Use Policy, 36*, 554–566.

LeCompte, M. D., & Schensul, J. J. (2013). *Analysis and interpretation of ethnographic data. A mixed methods approach.* AltaMira Press.

Lieskovský, J., Bezák, P., Špulerová, J., Lieskovský, T., Koleda, P., Dobrovodská, M., Bürgi, M., & Gimmi, U. (2015). The abandonment of traditional agricultural landscape in Slovakia: Analysis of extent and driving forces. *Journal of Rural Studies, 37*, 75–84.

Lindenfeld, L. A., Hall, D. M., McGreavy, B., Silka, L., & Hart, D. (2012). Creating a place for environmental communication research in sustainability science. *Environmental Communication: A Journal of Nature and Culture, 6*(1), 23–43.

Lukito, W., Malik, S., Surono, I., & Wahlqvist, M. (2015). From "lactose intolerance" to "lactose nutrition". *Asia Pacific Journal of Clinical Nutrition, 1*, 1–8.

MacDonald, D., Crabtree, J., Wiesinger, G., Dax, T., Stamou, N., Fleury, P., Gutierrez Lazpita, J., & Gibon, A. (2000). Agricultural abandonment in mountain areas of Europe: Environmental consequences and policy response. *Journal of Environmental Management, 59*(1), 47–69.

McDonnell, M. J., & Pickett, S. T. (1993). *Humans as components of ecosystems: The ecology of subtle human effects and populated areas.* Springer-Verlag.

Mládek, J., Kusendová, D., Marenčáková, J., Podolák, P., & Vaňo, B. (2006). *Demografická analýza Slovenska* [A demographic analysis of Slovakia]. Univerzita Komenského.

Murin, I., & Kandert, J. (2018). The daily rhythm of the villager of Central Slovakia: Comparative study of the chronological ethnography of four families from the Sihla plain in 1967–2017. *Český lid, 105*(1), 3–23.

Navarro, L., & Pereira, H. (2012). Rewilding abandoned landscapes in Europe. *Ecosystems, 15*, 900–912.

Novák, J., Podolák, J., Zuskinová, I., & Margetín, M. (2014). *Po stopách valachov v Karpatoch* [Following the footsteps of the Vlachs in the Carpathians]. Brno: Tribun EU.

Petrovič, F. (2006). Changes of the landscape with dispersed settlement. *Ekológia, 25*(1/2006), 201–211.

Priečko, M. (2015). *Socioekonomické perspektívy rozptýleného osídlenia na Slovensku (komparácia lokalit Málinec, Nesluša a Valaská Belá)* [Social and economic perspectives of dispersed settlement in Slovakia (based on the ethnological comparison of villages Málinec, Nesluša and Valaská Belá)]. Trnava: Univerzita sv. Cyrila a Metoda.

Raymond-Yakoubian, J., & Daniel, R. (2018). An Indigenous approach to ocean planning and policy in the Bering Strait region of Alaska. *Marine Policy, 97*, 101–108.

Ruddle, K. (1993). The transmission of traditional ecological knowledge. In J. Inglis (Ed.), *Traditional ecological knowledge: Concepts and cases* (pp. 17–32). Canadian Museum of Nature/International Development Research Centre.

Sandhu, H., & Wratten, S. (2013). Ecosystem services in farmland and cities. In S. Wratten, H. Sandhu, R. Cullen, & R. Costanza (Eds.), *Ecosystem services in agricultural and urban landscapes* (pp. 3–11). Wiley-Blackwell.

Schwarze, S. (2007). Environmental communication as a discipline of crisis. *Environmental Communication, 1*(1), 87–98.

Sjölander-Lindqvist, A., & Sandström, C. (2019). Shaking hands: Balancing tensions in the Swedish forested landscape. *Conservation and Society, 17*(4), 319–330.

Špulerová, J., Dobrovodská, M., Štefunková, D., Piscová, V., & Petrovič, F. (2016). History of the origin and development of the historical structures of traditional agricultural landscape. *Historický časopis, 64*(1), 109–126.

Steward, J. H. (1955). *Theory of culture change: The methodology of multilinear evolution*. University of Illinois Press.

Stoffle, R., & Minnis, J. (2008). Resilience at risk: Epistemological and social construction barriers to risk communication. *Journal of Risk Research, 11*, 55–68.

Stoffle, R. W., Stoffle, B. W., & Sjölander-Lindqvist, A. (2013). Contested time horizons. In A. Bond, A. Morrison-Saunders, & R. Howitt (Eds.), *Sustainability assessment: Pluralism, practice and progress* (pp. 51–67). Routledge.

Stoffle, R. W., Toupal, R., & Zedeno, N. (2003). Landscape, nature, and culture: A diachronic model of human-nature adaptations. In H. Selin (Ed.), *Nature across cultures: Views of nature and the environment in non-Western cultures* (pp. 97–114). Kluwer Academic Publishers.

Švecová, S. (1975). *Kopanicové sídla a dedina. Národopisná štúdia o spoločenských vzťahoch medzi obyvateľmi jednej slovenskej obce*. Univerzita Karlova.

Švorcová, J., & Kleisner, K. (2017). Jak s námi žije minulost: Epigenetika jako pojítko mezi kulturní a biologickou evolucí [How the past lives with us: Epigenetics as a link between cultural and biological evolution]. In L. Ovčáčková (Ed.), *O původu kultury. Biologické, antropologické a historické koncepce kulturní evoluce* [About the origin of culture: The biological, anthropological and historical concepts of cultural evolution] (pp. 349–371). Prague: Academia.

The Permanent European Conference for the Study of the Rural Landscape. (n.d.). http://www.pecsrl.org/index.html

Tindall, A. (1976). Theory in the study of cultural transmission. *Annual Review of Anthropology, 5*, 195–208.

Wheeler, H., Danielsen, F., Fid, M., Hausner, V., Horstkotte, T., Johnson, N., Lee, O., Mukherjee, N., Amos, A., Ashthorn, H., Ballari, Ø., Behe, C., Breton-Honeyman, K., Retter, G., Buschman, V., Jakobsen, P., Johnson, F., Lyberth, B., Parrott, J., Pogodaev, M., Sulyandziga, R., & Vronski, N. (2020). The need for transformative changes in the use of Indigenous knowledge along with science for environmental decision-making in the Arctic. *People and Nature, 2*, 544–556.

Community Voices, Practices, and Memories in Environmental Communication: Iliamna Lake Yup'ik Place Names, Alaska

Yoko Kugo

Introduction

In every culture and language, people use place names to refer to locations and to communicate information about these places with others. But how do people name a place, how do they identify the place in the landscape, and what is the significance of place names and stories about the places to the people? From an anthropological viewpoint, these questions prompt a researcher to consider how the use of language reflects understandings of landscapes through communication. Comprehending place names from insider (or community) perspectives contributes to the discipline of environmental communication by demonstrating that people share information about their environment through place names. Naming places and

Y. Kugo (✉)
Arctic and Northern Studies, University of Alaska Fairbanks,
Fairbanks, AK, USA
e-mail: ykugo@alaska.edu

A. Sjölander-Lindqvist et al. (eds.), *Anthropological Perspectives on Environmental Communication*, Palgrave Studies in Anthropology of Sustainability, https://doi.org/10.1007/978-3-030-78040-1_5

knowing these places supports residents' lifeways (traveling, fishing, hunting, gathering) and commemorates their family histories and wisdom people have passed down for many generations. People share their memories and oral traditions through place names and dialogues in their language and within their regions.

As in many Indigenous communities, Alaska Natives' cultural landscapes are inextricable from their identities. Cultural landscapes are a culmination of collective memory, personal experience, local histories, oral traditions, and place names that people have passed down within a cultural or regional group—this being one of many processes that aid in forming cultural identity (Andrews & Zoe, 1997; Cogos et al., 2017; Stoffle et al., 1997). The concept of cultural landscape enriches our understanding of environmental communication because cultural groups have practiced their land use in the region for many generations. For instance, the Yup'ik people (hereafter Yupiit, the plural for Yup'ik persons) of Southwest Alaska, who live in isolated areas and often in relatively small societies, have formed intimate relationships with the landscape through time by observing and participating in oral traditions with family and community elders (Fienup-Riordan, 1990, 2017). Many Yup'ik place names relate to the conditions of the land features, directional information, and the personal and collective memories of the community members and their relatives (Fienup-Riordan, 1990; Rearden & Fienup-Riordan, 2014; Salmon, 2014). These Yup'ik place names and narratives help the Yupiit define who they are as a people and as individuals, as well as their relationships to the environment.

In my previous fieldwork in Iliamna Lake communities, I discovered a cultural aspect of the people's relationship to the environment which I find crucial to environmental communication. Some Iliamna Lake Elders reported that when people harvest fish and animals, they thank these animals for giving themselves to people. To show respect, the people discard the bones of fish and animals in specific places in the water and in the ground (Kugo, 2014, pp. 180–181). Such oral traditions and practices concerning places involve an identity dimension, one which must be explored to gain a better understanding of Iliamna Lake Yup'ik environmental communication. This chapter explores how the study of place names contributes to environmental communication and anthropology, presenting stories of Iliamna Lake Yup'ik place names intertwined with their cultural ethics as the sustainability aspect of environmental communication. The term "environment" generally refers to earth, atmosphere, and physical locations. However, people often recognize environment as more than physical structures, such as the Southern Paiute people in North America, who see natural bridges as their ancestors' past (see Stoffle, chapter "Living Stone Bridges: Epistemological Divides in Heritage Environmental Communication" of this book). Many residents of Riddarhyttan in Sweden recognize that the land is contaminated, but wish to remain in the community because they have established relationships after living there for several generations (see Sjölander-Lindqvist, chapter "Arsenic Fields: Community Understandings of Risk, Place, and Landscape" of this book). Place names and stories of places prompt the Iliamna Lake residents to remember the landscape (physical locations) and their ancestors' voices and lessons (cultural practices), which are significant dimensions of environmental communication in the Iliamna Lake region.

A Study of Iliamna Lake Yup'ik Place Names

While building rapport with Iliamna Lake communities since 2012 and conducting literature reviews on this region, I learned that Iliamna Lake Yup'ik place names have rarely been recorded in the literature, in contrast with the over 200 Dena'ina place names in the Iliamna Lake region that have been published (Evanoff, 2010; Kari, 2013). Archaeological data and Russian explorers' records suggest that the language groups of Dena'ina Athabascan (Na Dené) and Central Yup'ik (Eskaleut) have lived in the region at least since the eighteenth century (Townsend, 1973; VanStone, 1988; Wrangell & VanStone, 1970). My dissertation project, a study of Iliamna Lake Yup'ik place names from 2016 to 2019, began with community members' request that I help them record and maintain Yup'ik place names and local histories about these places. The communities of Iliamna, Newhalen, Kokhanok, and Igiugig in the Iliamna Lake area and Levelock on the shore of the Kvichak River collaborated in the place names project that also served as my Ph.D. dissertation research. One Iliamna Lake resident told me that he wished they had begun this project when there were more Elders, "the original Googles," in the community (Iliamna Lake resident, pers. comm., Aug. 2016). Over 30 Elders, or those "original Googles," and community assistants in the five Iliamna Lake communities participated in eight ethnographic field trips between 2016 and 2019.

Iliamna Lake lies about 362 km southwest of Anchorage, in southwest Alaska. Today, five communities lie on the shore of the lake: Pedro Bay, Iliamna, Newhalen, Kokhanok, and Igiugig. Levelock, a community on the Kvichak River, has close ties to the Iliamna Lake communities. The residents of Levelock have long traveled to the Iliamna Lake communities to visit and exchange goods. The Kvichak River (approximately 120 km long) drains Iliamna Lake to Bristol Bay, which hosts the largest run of sockeye salmon returning to Iliamna Lake and its tributaries for spawning every summer. According to several Iliamna Lake Yup'ik Elders, their families came from the Kwethluk area on the Kuskokwim River to the present-day Newhalen area looking for food during a time of high starvation in the mid-nineteenth century (Coffing, 1991; Iliamna Lake Place Names Workshop, May 16, 2018). Owing to traveling, migration, and intermarriage, the Iliamna Lake communities include descendants of Dena'ina, Central Yup'ik, Alutiiq, Russian, and northern European people. Iliamna Lake Yup'ik Elders have heard that Russian and Euro-American explorers and surveyors borrowed or replaced original Yup'ik names and renamed them in Russian or English when they came to the area in the past. The recounting of the original Indigenous place names and stories about these places represents, according to Smith (1999), a decolonizing process. This is further illustrated by the fact that many Yup'ik place names and stories of places cannot be translated into other languages, or these translated names do not tell a deeper meaning of the names, especially stories about family histories and personal memories.

During the first year of my fieldwork, I discovered that some Iliamna Lake Yup'ik place names have multiple meanings or are embedded in stories that originated with individuals' experiences and memories. To gain insider perspectives for myself (an outsider) and to share local voices with residents, the village and tribal councils of Iliamna Lake communities and I established the Iliamna Lake Place Names Committee in December 2017. We held the Iliamna Lake Place Names Workshop in Newhalen in May 2018 to review many place names and locations of these places. As a result, our project recorded 219 Yup'ik and over 150 contemporary English names. To distribute our results to community members, we published two styles of place name maps, printing textile maps[1] for about 400

[1] "Place Names of Iliamna Lake, Nanvarpak, Nila Vena" (Nielsen et al., 2019).

residents and large paper maps[2] for research participants, local council offices, and schools. Iliamna Lake Yup'ik Elders, schools, and village/tribal councils obtained this large map for future projects and place-based education in their communities.

Place Names as Environmental Communication

People name places for various reasons, such as describing, honoring figures, commemorating historical events, narrating incidents, and adapting names through time, or their folk etymology. They share visual, directional, and historical information about these places through several forms of communication, including storytelling and participation in activities. Indigenous place names closely relate to insiders' use of language and oral narratives about these places within specific Indigenous cultures, from their own perspectives (Collignon, 2004; Cruikshank, 1990; Fair, 1997; Holton, 2011; Hunn, 2006; Kari, 1988; Ray, 1971; Rosaldo, 1980; Stewart, 1954; Thornton, 2008; Waterman, 1922). Such place names exemplify the Sapir-Whorf hypothesis that language, thoughts, and culture are deeply interconnected, and that the language people speak influences how they perceive their surroundings and conceive their own distinctive worldviews (Kluckhohn, 1961). Gumperz and Levinson (1996) reexamine the theory that such interconnected relationships with language differ among cultures and individuals, and that researchers should therefore address how language, thinking, and society are intertwined.

For instance, Basso (1988, 1996) discovered that Western Apache place names provide the people with mental advice, such as producing a mental image of a specific place, speaking about their ancestors, confirming their cultural values and morals, and/or offering comfort from being lost or stressed. The Western Apache also learn to "travel in [their] mind," so they can imagine the routes and feel confident about reaching these places (Basso, 1988, p. 123). While doing these tasks, Western Apache feel grateful for the help their ancestors' wisdom provides as they remember their ancestors and imagine their ancestors assisting them in traveling to the places. Western Apache place name practices represent environmental communication, as the Apache communicate with their ancestors, which help them to imagine the landscape physically and spiritually.

[2] "Place Names of Nanvarpak, Nila Vena, Iliamna Lake" (Kugo & Iliamna Lake Place Names Committee, 2019).

Kawagley's (2006) introduction to a Yup'ik worldview reflects environmental communication. The Yupiit not only communicate with the visible environment, but they also include the thoughts and feelings of humans, animals, and natural forces that represent parts of the surroundings. By sensing, hearing, and smelling the air outside, Yupiit become aware that *Ellam Yua* (the Spirit of the Universe) is watching over and providing for them to live on the land (Fienup-Riordan, 1990; Fienup-Riordan & Rearden, 2012; Kawagley, 2006). According to Kawagley (2006, p. 11), the Yupiit understand that "the land is a giver of life" by recognizing the oral tradition that the people were created and emerged in the Yukon-Kuskokwim region. In a Yup'ik worldview, people interact reciprocally with the natural or spiritual realms. In harvesting resources from their homeland, the Yupiit have a duty to care for the land. They fulfill this responsibility by taking care of harvested animals and deceased humans in ethical ways so that the spirits of the animals and people will return to the living world. The Yupiit discard the bones of harvested animals into the water or in the ground after they consume the meat and bone marrow. To show their gratitude to these people, the animals will be reborn and return to be killed by the same hunters. Fienup-Riordan (1990) describes how the Yupiit perceive "the relationship between humans and animals as collaborative reciprocity: the animals gave themselves to the hunter in response to his respectful treatment of them as persons in their own right" (p. 72). Such cultural ethics signify Yup'ik environmental communication in the Yukon-Kuskokwim region.

Researchers define the notion of environmental communication in multiple ways. Cox (2007) emphasizes preserving biodiversity as a central component of environmental communication that is vital in many cases. Environmental communication often lacks the spirituality and intimate relationships that local and Indigenous people have practiced in their homeland. Senecah (2007) points out that people being willing to engage each other about environmental concerns is essential to environmental communication. In her notion, environmental communication includes respect for local and Indigenous knowledge and the understanding of holistic ecological systems (Senecah, 2007, p. 28). Place names can be tools for such environmental communication because the knowledge they contain persists within the practices, memories, and social dimensions that community members use to nourish an intimate relationship with their environment (Cogos et al., 2017). This creates a sense of guardianship, because the land is a part of their community. In such cases, literal translations of Indigenous place names rarely describe a holistic community understanding of the environment. Yupiit did not name the Iliamna Lake Yup'ik stream *Quarruugvik* ("Place with Stickleback"), near present-day Igiugig, after its land feature of stream, or *kuik* (river). Rather, they named the place in recognition of their local knowledge of ecosystems, referring to stickleback (*Gasterosteus aculeatus*), a species of fish. People generally do not fish for small sticklebacks for food, but they know that the stream is a good feeding place for larger fish because of the presence of sticklebacks. Igiugig residents who live near the stream know that the mouth of the stream is a good place for ice fishing (pers. comm., May 2016). Local knowledge—knowing the abundance of fish in the stream and its Indigenous place name—and preserving this knowledge through communication assists the people in maintaining biodiversity.

In his essay "Environmental Communication: What It Is and Why It Matters," Mark Meisner (n.d.) defines "environmental communication" as "communication about environmental affairs." Meisner identifies two broad social functions of environmental communication: people use communication to do things, and communication shapes how people perceive and create meanings within their surroundings. Place names and stories about these places often offer people geographic information about the land, which allows them to practice their traditional lifeways and care for the land, which ultimately affirms the connections between the people and their ancestors who have passed down their wisdom to future generations.

Meisner (n.d.), and Stoffle, Arnold, and Bulletts (2016) also discuss how Indigenous people and government natural resource managers communicate with the environment differently. Stoffle, Arnold, and Bulletts (2016) explain that the Southern Paiute people perceive the world, plants, animals, and minerals as a *living universe* that is sentient, similar to humans. Scientists and natural resource managers rely on scientific evidence, such as changes in animal populations and mineral density in the soil, to manage environmental resources (Nadasdy 2005; Stoffle et al., 2016; Ross et al. 2011). In contrast, the Southern Paiute people have passed down their cultural ethics to communicate with plants, animals, crystals, and minerals to maintain a healthful environment (Stoffle et al., 2016). While environmental activists use communication power in furtherance of their environmental values, as they challenge government plans for creating dams, pipelines, and other land development for national economic growth (Harris, 2017), Indigenous communities, such as the Southern

Paiute and the Yukon-Kuskokwim Yupiit, communicate *with the environment* in fulfillment of their responsibility to maintain the relationships between animals, plants, land, spiritual beings, and people (Fienup-Riordan, 1990; Kawagley 2006; Stoffle et al., 2016). Environmental movements have emerged to protect the Earth's ecological system from human activity and global economy, but these movements typically reflect Western, rather than Indigenous orientations to the land. Yet Harris (2017, p. 77) points out that "local knowledge" is an effective means of environmental communication. As local knowledge is restricted to specific regions, Indigenous knowledge has been transmitted through an extended social system (e.g. from uncle to nephew) and practices as generations of Indigenous peoples have taught young people morals and ethics to live in harmony with their environments (Battiste & Henderson, 2000; Berkes & Folke, 1998). Similarly, place-naming practices are culturally and often locally unique, because they originate from oral information about personal and multi-generational experiences conveyed through native tongues and dialects. Collignon (2004) and Hunn (2006) call Indigenous place names "vernacular knowledge" that comprises observations of places and territories, and experiences in places, cumulatively shared by members of cultural groups. Indigenous place names, especially those in specific and regional languages, allow communication with the environment and the sharing of knowledge with limited groups—family and community members.

Colonizing and Decolonizing Iliamna Lake Names

The origins of Indigenous place names have often been obscured or reinterpreted as explorers and new settlers of lands have renamed places. After their arrival in Alaska in the mid-eighteenth century, Russian and Euro-American explorers recorded geographic information, including Native place names and new Russian and English place names, to expand trading routes with Alaska Natives and exploit natural resources on the land. Indigenous peoples orally shared their landscape information with others when traveling and telling stories about these places. For instance, Russian fur traders named Iliamna Lake in several ways, including *Oz[ero] Shelekhovo* (lit. "lake Shelekov") to honor a political figure[3] in 1786,

[3] Grigori Ivanovich Shelikov (Shelekov) was a founder of the Shelikhov-Golikov Company (later re-established as the Russian American Company), which first established a fur-trading

Ilyamna in 1796, *Lac Ilima* (lit. "big lake," its name in French) in 1827, and *Ozpo Ilima* and *Oz[ero] Bol[shoy] Ilyamna* (lit. "big Ilyamna lake" in Russian) in about 1850 (Orth, 1967, p. 449; Marvin Falk, email to the author, March 25, 2020). The Russian name *Ilyamna* is a distortion of the Dena'ina name *Nila Vena*.[4] Held by the Russian Military Archive, the map created by Russian explorers Izmailov and Bocharov, dated 1785–1786, is one of the earliest maps that labels Iliamna Lake as "lake Shelekov." The map in Fig. 1 shows Iliamna Lake's Indigenous names as *Nila Vena* ("Islands Lake") in Dena'ina and *Nanvarpak* ("Big Lake") in Yugcetun (Evanoff, 2010; Igiugig Village Council, 2012; Krauss et al., 2011).

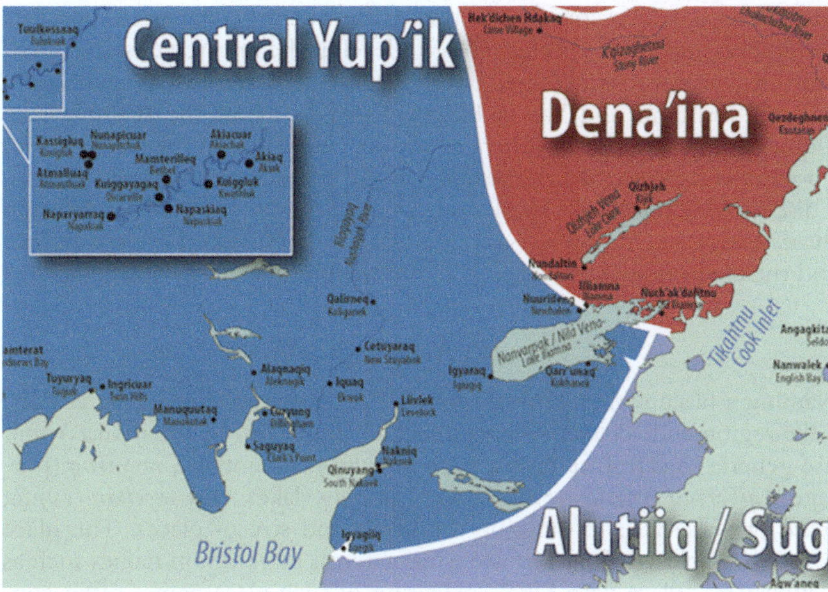

Fig. 1 Map of the area the author studied. Based on Krauss et al., 2011. (Image used courtesy of ANLC)

post on Kodiak Island around 1784. The Russian government later approved the RAC company's fur-trading monopoly, aiming to gain control over the land and people of Russian-America (Alaska) (Black, 2004, pp. 107, 255).

[4] According to some Iliamna Lake Elders, the present-day name "Iliamna" originated from the sound of its Dena'ina name, *Nila Vena*.

Iliamna Lake is the largest freshwater lake in Alaska, and it contains many small islands, especially in the northeast part of the lake, where the Dena'ina traditionally settled. Iliamna Lake Yup'ik Elders remember that some Dena'ina place names near the language boundary on the northeast side of the lake have been "Yupified," or borrowed and adapted to resemble Yup'ik phonology (Iliamna Lake Place Names Workshop, May 15, 2018). Recording and decolonizing Indigenous place names restores local histories and recognizes local observations and interactions with the landscape.

To better comprehend Iliamna Lake Yup'ik place names from insider perspectives, I used two methods: qualitative interviewing as used by oral historians, and a community-based participatory approach to gather ethnographic data. A community-based participatory approach prompts researchers and study communities to develop the research design collaboratively, thereby benefiting both parties. Working with community research assistants and Yup'ik Elders, I explored how the Iliamna Lake Yupiit communicate with others about the landscape. In the next section, I present some place names as examples of environmental communication. I discuss how the Yupiit speak and convey knowledge of places through these names and how they commemorate relationships between people and the environment, in the past and in the present.

Telling About Places

Naming a place with a general term is a common practice in the Iliamna Lake region and across other regions and languages. Iliamna Lake Yupiit use generic land feature terms, such as ***ingriq*** (mountain), ***tevyaraq*** (portage), ***qikertaq*** (island), ***kuik*** (river), ***nanvaq*** (lake), and ***igceńaq*** (falling place or waterfall), to describe the shape and size of places. The place names in Table 1 below also use generic terms or common names such as *Cuukvalek* ("Place with Pike"). Hearing of such place names often confused me initially, because I was not familiar with the landscape from their vantage points or memories, and therefore did not understand to which lake, river, or portage they referred to.

For instance, I recorded at least five *Cuukvaleks* and five place names that related to ***cuukvaq***,[5] or northern pike (*Esox Lucius*). One of the lakes

[5]Jacobson's Yup'ik dictionary defines ***cuukvak*** as "northern pike" (2012, p. 234). Northern pike (*Esox lucius*) is a species of carnivorous freshwater fish found in the Northern Hemisphere, including Alaska.

Table 1 Examples of Iliamna Lake place names, telling of places

Yup'ik name	*English translation*	*Narrators' descriptions of places (Ethnographic interview records compiled from my dissertation project between 2016 and 2019)*	*English name*	*Location*
Cuukvalek	"Place with Pike"	Pike fishing place	Schoolhouse Lake (named after the government built the school)	A lake near present-day Newhalen
Igceńaq	"Falling Place, Waterfall"	Falling place, historical accident at the place	Rapids	A fast-running stretch of the Newhalen River
Tevyaraq	"Portage"	Portage that goes from the lakeshore to the mountain	Roadhouse Mountain (named after a settler who built the roadhouse)	A mountain near present-day Iliamna
Amartetuli	"Customarily Packed"	Traditional seasonal campsite	Amakdedori (borrowing the Yup'ik name)	A historical site on the Kamishak Bay (a trail leads here from the Kokhanok area)
Tagyaracuar	"Small Place to Go Up"	Movement of walking on the trail	Pope Vannoy (named after the settler's surname)	A trail and site on the south side of Iliamna Lake
Kanaqlak	"Muskrat"	Traditional muskrat trapping place	Skater Lake (contemporary activity)	A lake near present-day Igiugig

(*continued*)

Table 1 (continued)

Yup'ik name	*English translation*	*Narrators' descriptions of places (Ethnographic interview records compiled from my dissertation project between 2016 and 2019)*	*English name*	*Location*
Cuukvagpalgem Kuiga, Olem Kuiga	"Big Pike's Creek","Ole's River"	Observed large pike, settler's cabin on the shore of the stream	Ole Creek (named after the settler)	A stream to the south of Iliamna Lake
Peksussurvik	"Place to Hunt Eggs"	Seasonal seagull hunting place	Egg Island (English translation of the Yup'ik name)	An island on the Kvichak River

named *Cuukvalek* lies near present-day Newhalen and is known in English as Schoolhouse Lake, since the government built a school near the lake in the early twentieth century. When I listened to the Elders' stories, I came to understand that the Yupiit named places after their geographic features (*Igceńaq* and *Tevyaraq*) and that stories about these places often convey meaningful information about these locations. The Yupiit also named places after their harvesting activities (*Cuukvalek, Kanaqlak, Cuukvagpalgem Kuiga*, and *Peksussurvik*), local histories (*Amartetuli* and *Tagyaracuar*), and observations of newcomers (*Olem Kuiga* and Pope Vannoy, the English name of *Tagyaracuar*). The English name Amakdedori originated from the sound of the Yup'ik name *Amartetuli* ("Customarily Packed"). *Amartetuli* is a historical settlement that sits on the shore of Kamishak Bay in Cook Inlet on Alaska's Southcentral Coast. People used to walk on multiple trails between the Kokhanok area, which lies roughly at the midpoint of the southern shore of Iliamna Lake, and *Amartetuli* (Amakdedori). Through stories, Yup'ik Elders remember the places where the people used to pack babies and supplies when they walked to *Amartetuli* in springtime, but this English name does not explain the meaning of the place. Stories of place names prompt the Iliamna Lake

Yupiit to remember historical events at these places and feelings toward the places and their locations.

Oral narratives of *Igceńaq* ("Falling Place" or "Waterfall"), one of the older Yup'ik settlements on the Newhalen River, demonstrate how Newhalen Yup'ik Elders understand the site as a fishing place, with Newhalen Yupiit family histories urging special caution because a person once drowned there (Fig. 2).

According to several Iliamna Lake Elders, their families who came from the Kuskokwim River to the Newhalen area settled and moved to three settlements prior to today's Newhalen location at the Newhalen River's mouth (Iliamna Lake Place Names Workshop, May 16, 2018; Kugo, 2014). Newhalen Yup'ik Elders remember the story of the Yupiit settling at and then abandoning a site along the Newhalen River called *Igceńaq* (Newhalen Elders, multiple personal communications, 2016, 2017, 2018). Newhalen Elder Annie Parks remembers that her father's family lived on the shore of *Igceńaq* before he married. She told of her father's oldest brother trying to spear a fish, and accidently falling into the rapids and being swept away. They never discovered his body.

Fig. 2 *Igceńaq* on the Newhalen River, May 2018. (Photo: Yoko Kugo)

> They used to put up fish [up there]. They don't fish [...] they don't seine or [...] just use *aggsuun* (harpoon) like this, then they catch fish. One day, [he] was fishing, that he happened to fall. *Igceńaq* [...] doesn't come back, just fall. [...] And they couldn't find a body. [...] We got to move out of there from that *Igceńaq*. (Annie Parks, interview, August 24, 2016)

Parks's parents were concerned about their future children falling into the rapids. For that reason, after this accident, they moved to another location near present-day Newhalen. While she was telling the story about *Igceńaq*, Parks continued to tell of other people moving downriver from the original *Igceńaq*. Fienup-Riordan explains that such sharing of personal knowledge, "either direct or transmitted by persons to whom living men and women can trace a relationship" refers to a *qanemciq*, a historical account in the local environment (Fienup-Riordan, 1990, pp. 103, 244). Telling a narrative about *Igceńaq* triggered Parks's memories of how her family lived in a harsh environment and their migration history.

When Parks tells a *qanemciq*, she always begins with a phrase "I only tell you what I know." This expression of hers emphasizes that she gained her knowledge about her family history and homeland by listening to stories from her relatives who witnessed events. She has established intimate relationships with her family and homeland throughout her lifetime. Parks showed me the trails where her grandmother walked to three lakes near the present-day Iliamna Airport to catch pike. Parks's story implies spatial information about the environment, the location of a Yup'ik settlement near *Igceńaq*, and temporal information that existed approximately three generations ago.

Basso (1996) uses the term "place-making" to refer to "a way of constructing history itself" (p. 6). Parks's story about *Igceńaq* exemplifies "place-making" in that Parks remembers her family history that she learned from her mother, as well as her own observation of the land. I did not record stories describing other waterfalls called *Igceńaq* as dangerous places; rather they are pleasant places to hike or swim. Parks's story about *Igceńaq* on the Newhalen River illustrates that stories of place names carry meaning to residents, in this case, to be aware of the danger at this great place for fishing.

Cultural Practices and Places

Telling of Yup'ik place names often prompted Elders to speak about their travel experiences and cultural ethics at these places. Iliamna Lake residents have transmitted such local knowledge through using place names in practices and activities such as navigation. According to Iliamna Lake Yup'ik Elders, many accidents occurred near *Qikertarpak* ("Big Island") in the southwest part of Iliamna Lake. When they tell stories about the accidents, they often include a story of gift-offering to the boulders, *Kass'aruayiit* ("Those That Look Like Priests"), which sit in the lake near *Ingrirpak* ("Big Mountain," Fig. 3). Some Iliamna Lake Yup'ik Elders reported that they make offerings by tossing food or pieces of crackers in the water when they go by *Kass'aruayiit*. By feeding *Kass'aruayiit*, the people will receive "good weather," according to Parks. Newhalen and Kokhanok residents must cross this area to reach the Kvichak River.

Fig. 3 *Ingrirpak* ("Big Mountain") viewed from a boat as the author and Iliamna Lake residents were crossing the southwest side of *Nanvarkak* ("Big Lake"), August 2018. (Photo: Yoko Kugo)

Conversely, the Igiugig people (who reside on the southwest of Iliamna Lake) remember these boulders as *Yuguat* ("Human-like, Fake Humans") because of their shapes. The Igiugig people commonly traveled by land along the shore to *Ingrirpak* ("Big Mountain"), but rarely by water. Therefore, Igiugig people did not perceive the boulders as an important landmark. The boulders' two names, originating from different stories, indicate that environmental communications vary depending on the people's physical and spiritual connections with the landscape.

Oral narratives that guide people to live respectfully with nature are ubiquitous in many societies, including those in southwest Alaska. The oral tradition of *An'gaqtar* (Stone Lady), in Togiak, Alaska, exemplifies how the Yupiit perceive gift-offering to the boulder Stone Lady to help them maintain a healthy reciprocal relationship between the human and animal worlds, as well as between the living and spiritual worlds. According to the story, *An'gaqtar*, a Yup'ik woman, lost her husband and failed to find food for herself and her child in the harsh environment. As a result, *An'gaqtar* lost her mind and went insane due to starvation, and she turned into the boulder (Fienup-Riordan & Rearden, 2012, pp. 48–50). Togiak Yupiit have rituals acknowledging *An'gaqtar* as their *Ellam Yua* (the Spirit of the Universe). They have practiced encircling *An'gaqtar*, moving in the sun's direction and giving offerings when they go by her. Fienup-Riordan and Rearden (2012) explain that the Togiak Yupiit understand *An'gaqtar*'s facing east as meaning that she welcomes both the sunrise and new lives that support peoples' survival: animals, fish, plants, or even a newborn baby in the family (p. 58). The Togiak Yupiit[6] beliefs surrounding *An'gaqtar* echo many Yup'ik creation stories about Raven creating rivers, mountains, and land for people and animals.

Iliamna Lake Yupiit do not recall how the boulders took the shape of *Kass'aruayiit*. However, their practice of sharing food with the *Kass'aruayiit* resembles that of the Togiak Yupiit making offerings to *An'gaqtar*. Both rituals present living environmental communications that reinforce their relationships with their homelands and maintain their environment's well-being.

While the Yupiit remember the landscape along with their cultural practices and oral narratives, government surveyors have recorded the land features and shorelines by measuring latitude and distance. Government surveyors' not having recorded these boulders' names suggests that they

[6] In this case, the form "Yupiit" refers to the possessor.

did not find them noteworthy. In their study of Sami place names, Cogos, Roué, and Roturier (2017) identify the government map publication process as "map making," which differs from "Sami mapping," wherein the Sami people use narratives to construct mental images of these places (p. 45). Narratives about *Kass'aruayiit* and *An'gaqtar* exemplify Cogos, Roué, and Roturier's (2017) notion of "micro-perspective," because these places hold regionally specific meanings to the residents. Place names can contribute critically to people's mental mapping as they travel, as in the case of *Qikertarpak* ("Big Island") and *Ingrirpak* ("Big Mountain") in the Iliamna Lake area. Government maps, or "map making," on the other hand, represent general environmental communication, presenting elevation, lengths of streams, and larger geographic features. Outsiders can use them to visualize potential natural resources around Iliamna Lake. In contrast, "Iliamna Lake Yup'ik mapping" exemplifies environmental communication from insider perspectives. Knowing how to travel safely on the large lake is crucial to Iliamna Lake residents for reaching their destinations and for harvesting seagull eggs, freshwater seals, and other resources around the lake.

Discussion

Narratives about place names and personal experiences at these places exemplify environmental communication from insider perspectives, because their interaction with the landscape comprises layers of memories from various times and accumulated residents' knowledge. Their conversations about place names and stories about these places convey and reinforce the significance of the place names to the people in the past and present. When Iliamna Lake Yup'ik Elders talked about place names, they typically started telling a story about where they grew up and where they used to travel with family from their settlement to another settlement. Their storytelling reflects environmental communication in connecting their family histories, way of life, and memories of the land features with their images of their landscape. Travel routes connecting the place names to one another demonstrate how the Yupiit lived, moved throughout the year harvesting food, and stayed at the fish camp on the shore of the river in the summer (Fienup-Riordan, 1990, p. 9). Ingold (2011) calls such geographic knowledge of the landscape a "web of knowledge," explaining that places are like knots in the web, and multiple travel routes connect those knots. For instance, some Yup'ik place names associated with fish camp, pike, and suckerfish in the Iliamna Lake area indicate that the Yupiit observed the abundance of fish and harvested them. A small group of Yupiit settled near *Igceńaq* because it lies on the shore of the river and is also within walking distance to small lakes where they can harvest pike.

Kawagley's explanation of the Yup'ik worldview and Basso's study of Western Apache place names show that Yupiit and Western Apache environmental communications are regionally specific and transmitted through dialogues and other practices. Iliamna Lake residents' understanding of some place names demonstrates their concerns about the environment. Residents refer not only to place names as physical landmarks, but they also strengthen spiritual and cultural connections to the land. Some Newhalen and Kokhanok residents believe that offering gifts to the boulders *Kass'aruayiit* ("Those That Look Like Priests") will ensure that the *Kass'aruayiit* watch over the peoples' safe passage. Residents remember an accident that happened at *Igceńaq*, whereas government land surveyors might simply record *Igceńaq* as "Rapids." Local knowledge embedded in the place name provides an environmental alert for the people as they move about the land. Senecah (2007) predicts that "collaborative processes will define the future of much environmental decision making, from local to global levels" (p. 28). Understanding place names and histories from insider perspectives can provide valuable insight in such decision making and promote the longevity of community prosperity.

The gift-offering practice reflects a core value among Alaska Natives. Sharing fish, meat, and plants with others and showing thankfulness to animals, landscape, and spiritual beings shows reciprocity and ensures future catches and safe travels. As Fienup-Riordan (2005) observes, being thankful to one another is "ubiquitous in Yup'ik daily life" (p. 59). This cultural ethic of offering food to *Kass'aruayiit* is a mnemonic device, reminding the people of accidents that occurred near there in the past and ensuring safe journeys across the lake. Iliamna Lake people do not need scientific proof of the power of giving gifts to *Kass'aruayiit*. Rather they do it because "it has been said"; their Elders have taught them to do so. According to Yup'ik professor Walkie Charles, a Yup'ik person does not own a narrative, story, or information about life in general. These stories and teachings are passed on from one person to another. In the old days, spoken Yugcetun was the medium of discourse; there was no way to write events or stories down on paper. Like many Indigenous languages, Yugcetun was initially a spoken, unwritten language until first contact by explorers, educators, and clergy (Barnum, 1901; Hinz, 1944; Nelson,

1983). Storytelling and oral sharing of teachings and learning were the basis by which to make sense of the world around them for the survival of language and culture (Walkie Charles, personal communication, July 12, 2020). Stories containing cultural ethics are intertwined with place names and remind community members to be conscious of the environment.

When sharing physical information and histories of places, such place names perform an environmental communication role, in effect, participating in activities, shaping peoples' memories, and creating meanings within the landscape, as Meisner (n.d.) explains. Many Yup'ik place names require language skills to understand the deeper meanings of the names that represent an important part of Indigenous knowledge. Colonized and contemporary English names do not contain such cultural meanings, but they show the arrival of newcomers to the land, the modern economy, and the dominance of the English language through time. In using local place names and relating narratives about them, Iliamna Lake Yup'ik Elders share their wisdom and feelings of the land and tell their stories from their insider perspectives. In this way, recording and learning stories of Iliamna Lake Yup'ik place names decolonizes, and this sense of responsibility to the environment can help preserve biodiversity and sustainability.

Researchers can gain insight into environmental communication through ethnographic fieldwork and learning local usage of Indigenous place names with community residents. Iliamna Lake Yup'ik place names like *Quarruugvik* ("Place for Sticklebacks"), *Igceñaq* ("Falling Place"), and *Kass'aruayiit* ("Those That Look Like Priests") exemplify important roles in regionally specific environmental communication. I hope that the study of Iliamna Lake Yup'ik place names affirms the residents' cultural pride and that they continue to maintain their lifeways. Sharing place names and stories about these places is one way for the Iliamna Lake residents to communicate with others about the environment and to live in harmony with the world.

Acknowledgments I would like to thank the University of Alaska Fairbanks Center for Global Change and National Science Foundation (No. 1640812) for supporting my dissertation project. The Bristol Bay Native Corporation Education Foundation supported the Iliamna Lake communities' place names projects that I collaborated with on my dissertation. I also extend my thanks to the Iliamna Lake Place Names Committee and research participants who shared their time and stories with me.

References

Andrews, T. D., & Zoe, J. B. (1997). The *Įdaà* trail: Archaeology and the Dogrib culturallandscape,NorthwestTerritories,Canada.InG.P.Nicholas&T.D.Andrews (Eds.), *At a crossroads: Archaeology and First Peoples in Canada* (pp. 160–177). Archaeology Press Department of Archaeology Simon Fraser University.

Barnum, F. S. J. (1901). *Grammatical fundamentals of the Inuit language: As spoken by the Eskimo of the Western Coast of Alaska*. Ginn & Company, Publishers.

Basso, K. H. (1988). "Speaking with names": Language and landscape among the Western Apache. *Cultural Anthropology, 3*(2), 99–130.

Basso, K. H. (1996). *Wisdom sits in places: Landscape and language among the Western Apache*. University of New Mexico Press.

Battiste, M. A., & Henderson, J. Y. (2000). *Protecting Indigenous knowledge and heritage: A global challenge*. Purich Pub.

Berkes, F., & Folke, C. (1998). Linking social and ecological systems for resilience and sustainability. In F. Berkes, C. Folke, & J. Colding (Eds.), *Linking social and ecological systems: Management practices and social mechanisms for building resilience* (pp. 1–25). Cambridge University Press.

Black, L. (2004). *Russians in Alaska: 1732–1867*. University of Alaska Fairbanks.

Coffing, M. (1991). *Kwethluk subsistence: Contemporary land use patterns, wild resource harvest and use, and the subsistence economy of a Lower Kuskokwim River area community*. Alaska Department of Fish and Game, Division of Subsistence.

Cogos, S., Roué, M., & Roturier, S. (2017). Sami place names and maps: Transmitting knowledge of a cultural landscape in contemporary contexts. *Arctic, Antarctic, and Alpine Research, 49*(1), 43–51.

Collignon, B. (2004). It's a long way to the other geographers and geographic knowledges. *GeoJournal, 60*(4), 375–379.

Cox, R. (2007). Nature's "crisis disciplines": Does environmental communication have an ethical duty? *Environmental Communication, 1*(1), 5–20.

Cruikshank, J. (1990). *Life lived like a story: Life stories of three Yukon Native Elders*. University of Nebraska Press.

Evanoff, K. (2010). *Dena'ina Ełnena, a celebration: Voice of the Dena'ina*. National Park Service Lake Clark National Park Preserve.

Fair, S. W. (1997). Inupiat naming and community history: The Tapqaq and Saniniq coasts near Shishmaref, Alaska. *The Professional Geographer, 49*(4), 466–480.

Fienup-Riordan, A. (1990). *Eskimo essays: Yup'ik lives and how we see them*. Rutgers University Press.

Fienup-Riordan, A. (2005). *Wise words of the Yup'ik People: We talk to you because we love you*. University of Nebraska Press.

Fienup-Riordan, A. (2017). *Qanemcit Amllertut/Many stories to tell: Traditional tales and narratives from Southwest Alaska* (A. Rearden & M. Meade, Trans.). Fairbanks: University of Alaska Press.

Fienup-Riordan, A., & Rearden, A. (2012). *Ellavut/Our Yup'ik world and weather: Continuity and change on the Bering Sea coast.* University of Washington Press.

Gumperz, J., & Levinson, S. C. (1996). Introduction to Part I. In J. Gumperz & S. C. Levinson (Eds.), *Rethinking linguistic relativity* (pp. 21–36). Cambridge University Press.

Harris, U. S. (2017). Engaging communities in environmental communication. *Pacific Journalism Review, 23*(1), 65–79.

Hinz, J. (1944). *Grammar and vocabulary of the Eskimo language: As spoken by the Kuskokwim and Southwest Coast Eskimos of Alaska*. The Society for Propagating the Gospel, the Moravian Church.

Holton, G. (2011). Differing conceptualizations of the same landscape. In D. M. Mark (Ed.), *Landscape in language: Transdisciplinary perspectives* (pp. 225–237). John Benjamins Publishing Company.

Hunn, E. S. (2006). Meeting of minds: How do we share our appreciation of traditional environmental knowledge? *Royal Anthropological Institute*, 143–160.

Igiugig Village Council. (2012). *Our story, our people, our home.* Igiugig Village Council.

Ingold, T. (2011). *Being alive: Essays on movement, knowledge and description.* Routledge.

Izmailov, G. G., & Bocharov, D. I. (1785–1786). *Map of the Alaska Peninsula.*

Jacobson, S. A. (2012). *Yup'ik Eskimo dictionary (Vols. 1 & 2).* Alaska Native Language Center, University of Alaska Fairbanks.

Kari, J. M. (1988). Some linguistic insights into Dena'ina prehistory. *Alaska Anthropological Association Monograph Series, 4*, 319–338.

Kari, J. M. (2013). *Dena'ina topical dictionary.* Alaska Native Language Center.

Kawagley, A. O. (2006). *A Yupiaq worldview: A pathway to ecology and spirit* (2nd ed.). Waveland Press.

Krauss, M., Holton, G., Kerr, J., & West, C. T. (2011). *Indigenous peoples and languages of Alaska* [map]. Fairbanks: Alaska Native Language Center and UAA (University of Alaska Anchorage) Institute of Social and Economic Research. Retrieved from https://www.uaf.edu/anla/map

Kugo, Y. (2014). *Subsistence practices of Iliamna Lake Villages: An investigation of dynamics of traditional and local ecological knowledge.* M.A. thesis, University of Alaska Anchorage, Anchorage.

Kugo, Y., & Iliamna Lake Place Names Committee. (2019). *Place names of* Nanvarpak, Nila Vena, *Iliamna Lake* [map]. Fairbanks: Alaska Native Language Center.

Kluckhohn, C. (1961). Notes on Some Anthropological Aspects of Communication. *American Anthropologist, 63*(5), 895.

Meisner, M. (n.d.). *Environmental communication: What it is and why it matters.* International Environmental Communication Association. Retrieved June 17, 2018, from http://theieca.org/resources/environmental-communication-what-it-and-why-it-matters

Nadasdy, P. (2005). The anti-politics of TEK: The institutionalization of co-management discourse and practice. *Anthropologica, 47*(2), 215–232.

Nelson, E. W. (1983). *The Eskimo about Bering Strait. Vol. 18, pt. 1, Bureau of American Ethnology annual report for 1896–97.* Smithsonian Institution Press. (Original work published 1899).

Nielsen, M., Kugo, Y., & Iliamna Lake Place Names Committee. (2019). *Place names of Iliamna Lake*, Nanvarpak, Nila Vena [map]. Kokhanok: Kokhanok Village Council.

Orth, D. J. (1967). *Dictionary of Alaska place names* (revised 1971). Washington, DC: Government Printing Office.

Ray, D. J. (1971). Eskimo place-names in Bering Strait and vicinity. *Names, 19*(1), 1–33.

Rearden, A., & Fienup-Riordan, A. (2014). *Nunamta Ellamta-llu Ayugucia: What our land and world are like: Lower Yukon history and oral traditions.* Alaska Native Language Center.

Rosaldo, R. (1980). *Ilongot headhunting, 1883–1974: A study in society and history.* Stanford University Press.

Ross, A., Sherman, R., Snodgrass, J. G., & Delcore, H. D. (2011). *Indigenous peoples and the collaborative stewardship of nature: Knowledge binds and institutional conflicts.* Left Coast Press.

Salmon, A. (2014). The Kukaklek Reindeer Station. *Alaska Journal of Anthropology, 12*(2), 46–51.

Senecah, S. L. (2007). Response to Cox: Impetus, mission, and future of the environmental communication commission/division: Are we still on track? Were we ever? *Environmental Communication, 1*(1), 21–33.

Smith, L. T. (1999). *Decolonizing methodologies: Research and indigenous peoples.* Zed Books.

Stewart, G. R. (1954). A classification of place names. *Names: A Journal of Onomastics, 2*(1), 1–13.

Stoffle, R. W., Arnold, R., & Bulletts, A. (2016). Talking with nature: Southern Paiute epistemology and the double hermeneutic with a living planet. In G. Tully & M. Ridges (Eds.), *Collaborative heritage management* (pp. 75–99). Gorgias Press.

Stoffle, R. W., Halmo, D. B., & Austin, D. E. (1997). Cultural Landscapes and Traditional Cultural Properties: A Southern Paiute View of the Grand Canyon and Colorado River. *American Indian Quarterly, 21*(2), 229–249.

Thornton, T. F. (2008). *Being and place among the Tlingit*. University of Washington Press.

Townsend, B. J. (1973). Ethnoarchaeology in nineteenth century southern and western Alaska: An interpretive model. *Ethnohistory, 20*(4), 393–412.

VanStone, J. W. (1988). *Russian exploration in Southwest Alaska: The travel journals of Petr Korsakovskiy (1818) and Ivan Ya. Vasilev (1829)*. University of Alaska Press.

Waterman, T. T. (1922). The geographical names used by the Indians of the Pacific Coast. *Geographical Review, 12*(2), 175–194.

Wrangell, F. P. V., & VanStone, J. W. (1970). The inhabitants of the Northwest Coast of America. *Arctic Anthropology, 6*(2), 5–20.

Demographic Change and Local Community Sustainability: Heritagization of Land Abandonment Symbols

Ivan Murin, Jan Horský, and Ján Aláč

Introduction

In this chapter we bring the reader to the village and municipality of Horný Tisovník in the south of Central Slovakia. It was here, from 2016 to 2020, that a number of anthropologists and their students worked with the people of this and other villages to understand their connections with the land, the community, and each other. These connections were reflected in the local culture. As local populations disappeared in the twentieth century, the local forms of arable land, settlements, terraced fields, grasslands,

I. Murin (✉) • J. Aláč
Faculty of Humanities, Charles University, Prague, Czech Republic
e-mail: ivan.murin@umb.sk

J. Horský
Faculty of Arts, Charles University, Prague, Czech Republic
e-mail: honza.horsky@seznam.cz

A. Sjölander-Lindqvist et al. (eds.), *Anthropological Perspectives on Environmental Communication*, Palgrave Studies in Anthropology of Sustainability, https://doi.org/10.1007/978-3-030-78040-1_6

and orchards quickly vanished along with them. Some of their remnants, such as walkways, roadside marks and crosses, and cemeteries, which we refer to as "land abandonment symbols," are a foundation for the ***heritagization*** of this and similar communities. Here we argue that the depopulation of large areas and subsequent changes in landscape and community culture can be reversed by turning to traditional eco-social activities and family farming as a foundation for establishing more sustainable communities.

The sustainability of communities like Horný Tisovník began to deteriorate in the first half of the twentieth century, when there was a dramatic reduction in birth rates. As a consequence, the local traditional communities began to experience a decline in population, which in turn caused extensive environmental and social changes. Because there were fewer community members, beneficial activities were neglected. In the short term, the shares of land ownership for individual families increased, benefiting innumerable families with one child. Later, with the advancing loss of workforce, only the most viable land was cultivated, in the end almost none. All the energies of the declining communities were focused securing their livelihoods. The smaller communities could not perform the activities necessary for the smooth functioning of the local government and the maintenance of roads, public spaces, and buildings. The only solution seemed to be to rent or sell land. Respect for local authorities, undermined by loss of vitality and depopulation, declined. Much farmland could not be managed, and so saltbush returned to places that had traditionally been open landscapes.

This historical example of a local community's decline is one of many in recent European history. Its impacts on the landscape, however, are only now being recognized. By following local cultural and natural trajectories of vitality, anthropological findings constitute a salient narrative in environmental discussions. Analyses of complex local culture–nature connections offer an alternative type of support for argumentation on global issues. In environmental communication, translocal elements are thus significantly represented, creating a new argumentative space for predicting future impacts.

The Depopulation of Cultural Regions in Europe

Although the world population is experiencing constant growth, concerns related to the loss of diversity of residential areas and culture, as well as the loss of vitality of populations in individual regions around Europe (The Permanent European Conference for the Study of the Rural Landscape, 2020) are gaining priority in public and academic space. Depopulation of certain areas has multiple causes, which are often mutually connected, and that can lead to serious consequences. One trend is that particularly young generations of productive and reproductive age leave for urban and suburban areas, and while this migration generally causes dynamic growth in certain conurbations, it nevertheless has a disturbing regressive effect on rural residential areas. It is mainly the countryside, or the village areas of certain regions, which are most affected by depopulation (Gajdoš, 2016). Populations which have formed distinctive cultural configurations have been declining and dying out. In relation to other problems, population decline creates a vicious cycle known as *circular and cumulative causation* (Moravčíková & Fürjészová, 2018), which is very difficult to break (Massey, 1990). The main structure of cultural configurations is easily replicated without enriching selections and innovations of individual elements. New, equally stagnant structures are bundled into non-adaptive strategies. From an anthropological and ethnological point of view, the process of depopulation plays an important role in cultural diversity loss. Long-existing strategies and rich contexts of multigenerational adaptation to specific past and future environments accumulate and contain important material and spiritual cultural messages.

The future of the cultural landscape should be perceived as linked to current global challenges, and this holds a significant place in contemporary anthropology's research agenda (Stoffle et al., 2020). Among the main challenges facing rural areas in the twenty-first century are climate change, migration, population aging, depopulation, technological innovations, and urbanization. Rural areas are heterogeneous, and it is important to realize that some rural settlements are located near urban centers, within larger integrated areas, while others are extremely distant and often have to fight for survival. Life in the countryside and cultural identity are closely connected, and it is clear that various changes have an impact on local rural communities, which leads to them taking irregular development trajectories (Murin & Kandert, 2018).

Background

Small local communities in Central Europe have always played an important role in the transformation of cultural landscapes, while their extinction is closely connected with concerns about whether cultural and natural identity in Europe is sustainable. From the anthropological point of view, the link between the culture of local communities and the environment can be seen within the complicated processes of cultural adaptation. It is a paradigm of anthropology that adaptation processes are continuous, and thus form cultural mechanisms maintaining the sustainability of both humankind and the environment. Dialogues on the topic often speak to the need to define our environment's human-carrying capacity. The concerns regarding the growth of the human population are considerable. Slightly less do we encounter studies that deal with environmental problems with the opposite cause, that is, those that concern the depopulation of landscape. There is a consensus among anthropologists that people constitute a land-forming factor, and human populations diversify culture and nature by way of their long-term activities. Environmentalists and anthropologists give numerous examples of the optimal coexistence of humans and nature. In a case study based in south-central Slovakia, examples can be found of degradation of the landscape and the local community (Ustaoglu & Collier, 2018) caused by the depopulation of local communities (Westhoek et al., 2006) through consciously reduced birth rates among women (Andorka et al., 1998). This has been a reaction—a maladaptive human behavior (Boyd & Richerson, 2005; Lawson & Borgerhoff Mulder, 2016; Sear et al., 2016)—to reduced work opportunities in the traditional agricultural landscape (Špulerová et al., 2016).

Local Communities and the Cultural Landscape

Geographically, locally distinctive rural communities in Central Europe are not linked to larger geographical units. Their relative seclusion during their development has meant that for a long time they have functioned as a somewhat closed distributive model of their own culture. The relatively common physical and cultural environment was shared, a place where shared cultural meanings were selected, passed on, and accepted. Local culture is, therefore, based on real, adopted meanings, which are, to some extent, understood by all its members. Appadurai (1986) describes performance, representation, and action as the means through which people socialize the landscape. He proposes the term "ethnoscape" (Appadurai, 1996, p. 64) to refer to villages, communities, localities, societies, and so on.

The knowledge of each local community is documented in the local collective, individual, and these days also technological, memory, which represents the community's "database." This knowledge's distribution depends on changes relevant to the community's location. If a change in community localization occurs, for example, the population moving out or decreasing, the decoding of local meanings decreases. Without interaction, direct contact with the environment, social learning, and generational transmission, the symbolic meanings that relate to individual cultural phenomena and information cannot be grasped correctly. Those that are shared indirectly and acquired by new experience are highly innovative but culturally less stable. Symbolic meanings of culture emerge from the thinking, actions, and communication of an optimally populated group of people. This logic of reproductive behavior develops in close interaction with the time and stability of the environment.

Due to depopulation, these ordinary ways of being in the landscape (Ingold, 2000), the building of a sense of locality (or knowing of the landscape) has been interrupted.

> Central among these facts is the changing social, territorial, and cultural reproduction of group identity. As groups migrate, regroup in new locations, reconstruct their histories, and reconfigure their ethnic projects, the ethno in ethnography takes on a slippery, nonlocalized quality, to which the descriptive practices of anthropology will have to respond. (Appadurai, 1996, p. 48)

The degradation of the landscape "erases" the supporting information of the cultural memory in local communities. Ethnoscapes, as places of community identity, are deterritorialized (Appadurai, 1996, p. 52) into barely informative units without the lived experience of humanity.

The Historical Context of Central European Migration and Depopulation

If we define historical "Central Europe" as the area bounded by the Alps, the Rhine, the Baltic Sea, and the borders between the West and East of Christianity, the migration processes from the sixteenth century to the present can be typologically classified as follows (Kárníková, 1965):

1. Migration (population transfers) associated with wars and political changes. The largest of these are mostly (ethnically motivated) one-off movements of millions of inhabitants associated with the end of World War II. Other such cases of migration include more gradual processes, such as state-organized relocation of populations to open spaces.
2. Urban migration in the second half of the nineteenth and the twentieth centuries to large urban centers (Vienna, Buda-Pest) and industrial city centers. Despite the rapid population growth from the eighteenth to the first half of the twentieth centuries, this migration led, though only in some areas, to the depopulation of certain rural regions (Zeitlhofer, 2006, p. 127).
3. Emigration from the poorer regions of Central Europe to North America.
4. Steady economic migration from eastern to western Central Europe. In the past 30 years, Central European countries have been the beneficiaries of not only unskilled labor via east–west migration but also doctors and other medical staff.
5. Frequent, abundant seasonal economic migration is well established throughout Central Europe.
6. Long-term small-scale migration within individual regions has long been abundant. For example, there is evidence of marital migration reaching as far back as the sixteenth century (Grulich & Zeitlhofer, 1998).
7. Depopulation and related migration caused by a conscious reduction in reproductive behavior.

The character and dynamics of these processes vary locally, depending on the economic character of the particular Central European region, with respect to inheritance law and their prevailing cultural type, or ***marriage pattern***. When distinguishing cultural types in Central Europe, anthropologists use Hajnal's Saint Petersburg–Trieste line (Hajnal, 1965), which corresponds (approximately) to the historical internal border between Austria and Hungary. Northwest of this line, local communities strictly adhered to the principle of neolocalism (where marriage meant a new household, i.e., a new "locality"), or to the principles of inheritance law. In the countryside, the preferred practice was impartible inheritance, that is, inheritance by only one heir, who then paid the other heirs a sum equivalent to their share of the estate (Laslett & Wall, 1972). Southeast of the Hajnal line, marriage was not so consistently linked to obtaining the existing locality or building a new estate, and therefore the land was divided and the area was populated more densely.

Land Depopulation: Limited Fertility as a Maladaptation

One of the most significant features of modern European culture is the decline in marital fertility. The fertility ratio has decreased from its traditional rates in two transformational waves. The traces of Europe's first, "big transformation" (Livi-Bacci, 2000, pp. 154–198) could be observed in the mid-eighteenth century, continuing through the nineteenth and the first decades of the twentieth centuries. This first demographic transition was caused by a decrease in the death rate, particularly that of babies and children. During this transition the population rose quickly, which in turn led to a subsequent use of birth control and lower birth rates for married couples.[1]

The second demographic transformation was observed in the 1950s, becoming more apparent in the 1960s (Lesthaeghe, 2010; Esping-Andersen & Billari, 2015). In the last quarter of the twentieth century, this transformation significantly impacted Central Europe (Pavlík, 2018).[2] Those local communities of Central European countries that had seen sustained population growth in the previous two centuries began to stagnate and age, due to the significant decline of the fertility/birth rate, which manifested itself in the growing percentage of elderly inhabitants in the overall population.

[1] In France in 1900, for example, one woman had 2.79 children on average, while in other European countries, this average was still between 3 and 5 per woman.

[2] The average number of children per woman fell deeply to under 2; in Italy of 1995 it was 1.17.

The Increasing Control of Birthrate Among Married Couples

The area of Central Europe that most exemplifies the impact of the strictly upheld one-child practice on the depopulation of land is the area stretching along the northern banks of the Dráva and Danube rivers. The westernmost such area is the Somogy region of Hungary, south of Lake Balaton. Approximately 300 km to the east, the Krassó-Szörény region represents the end of the one-child area (Vasary, 1989). The northern parts of this area are comprised by the southern Slovak Carpathians. The whole area is now divided among Hungary, Slovakia, Serbia, and Romania. At the turn of the nineteenth and twentieth centuries, when the one-child practice reached its peak, this area had approximately three million inhabitants and was mainly characterized by a predominantly agrarian way of life (Buday, 1909). The one-child norm did not arise from social and economic modernization in the area, but rather from the inability to establish independent households, especially outside the agrarian systems.

Fertility limitations at the turn of the nineteenth and twentieth centuries were first accepted on an individual level, later spreading throughout the stabilized restrictive social norms of the local community (Botiková, 2016). The land properties of the individual families were too small, hovering around the subsistence level in certain types of settlements (Holec, 2003, p. 222). At first postponing the birth of the first child prevailed, and only later was there a decrease in the total number of children. Reduction of the population caused an increase in farms' revenues in a very short period of time.

The one-child practice contributed to the accumulation of land ownership, and very quickly—over the course of two or three generations in the early twentieth century (Fig. 1) the material benefits of having one child became clearly apparent. Due to the consistency with which the one-child practice was adopted, as well as due to the local endogamy, the land was merged into larger units, owned by a decreasing number of families. Arrangement of potential marital partnerships was done from a very early age, and it was not always successful. Consequently, there was a general lack of single men, while young women remained single and without children. This led to fundamental changes in the localities' social climate. Formerly hardworking people and modest farmers began to succumb to excessive consumerism. The farmers, whose parents and grandparents lived in small traditional dwellings were building large, monumental

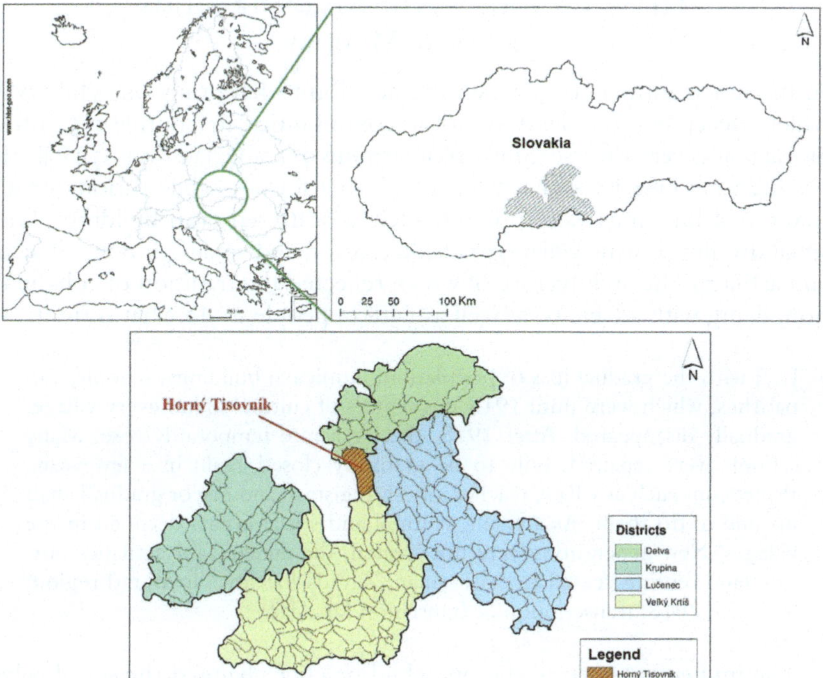

Fig. 1 The spread of the one-child practice in regions of south-central Slovakia (maps above) and the location of the Horný Tisovník research locality (map below). (This image used with permission of Pavol Midula)

houses in order to impress their neighbors and peers, and losing their vitality in the process.

Later, after approximately five generations of reduced birth rates, the local communities began to experience a disproportionate fall in their numbers, which reflected the emptying of remote places in the land. The reduced population started to neglect communal activities that were beneficial to the locality. They avoided participating in those activities that were necessary for the smooth functioning of the local government, road maintenance, and maintenance of public spaces and buildings.

Heritagization as the Reconstruction of Local Memory

In the nineteenth century, when the decision to have only one child was widely accepted, the land of local communities in southern Central Slovakia was very diverse. Small-plot farming of arable land was typical—it has high biodiversity and a diversity of relief forms—while various small, traditional buildings were present, such as scattered farm buildings, seasonal dwellings, wine cellars, roadside crosses, and small symbolic objects and artifacts, the architecture of which reflected the tradition of cohabitation, living with adversity, as well as faith in people to do what is right.

> [...] with the gradual loss of population, municipal buildings, schools, and parishes, which were until 1918 the centers of cultural life in every village, gradually disappeared. After 1918, the birth rate temporarily rose. Many schools were repaired, only to be gradually closed again in a few years. Professions such as village teacher, priest, registrar, and mayor gradually had no one to do them. As a result, cultural and social events declined in the villages. New generations were not created, the demand for activities outside land renting ceased to exist. (man, Závada locality, Novohrad region, Slovakia, 2018, transcription & translation Ján Aláč)

The implementation of the one-child practice changed the social roles of the local community members. The eldest women became most important, their dominance hinging upon the accepted norms of reproduction, as they were the main guardians of the one-child practice in families. If daughters or daughters-in-law became pregnant, it was mostly the eldest women who made decisions regarding the unborn. Their roles grew in importance, especially when the second child could endanger the wealth prospects of new families. Supported by other women, the eldest shaped and influenced the opinions within the local communities. These continued reductive habits and accumulation of inherited farmland among a small number of heirs contributed to the continuous degradation of traditional farming.

> If a single child has left for a better life, work, education or self-realization, he hasn't returned to his parents' farm. Many dwellings in the village have remained abandoned and redundant. Here in our villages, where a one-child custom has been adopted, we still have a high number of abandoned, rundown homes and entire farms compared to other regions. (woman, Závada

locality, Novohrad region, Slovakia, 2018, transcription & translation Ján Aláč)

In general, during Slovakia's massive industrialization in the second half of the twentieth century, family farming was replaced by state-run large-scale agricultural production. New generations did not acquire the virtue of sharing the land or maintaining an austere way of life, nor did they learn to understand the role of sociability, which was essential to traditional farming activities. The offspring did not see any reason to remain in the rapidly declining peripheries, and began to move into cities, or to areas of developed industrial production.

Historical Cemeteries and Tombstones: Places of Cultural Memory

In the south-central Slovak regions of Hont and Novohrad, field research has been conducted regarding the impact of one-child practice on local cultures.[3] Since 2008, along with the inhabitants of the Horný Tisovník locality, we have been trying to revive the local cemeteries, which, in their symbolic beauty, also carry the painful legacy of the collective memory inherited by the remaining local population, which does not exceed 180 today (in 1910, the local population was 1464). As the decrease in the permanent population continues, one of the ways to stabilize it is by creating an eco-museum focusing on revitalizing traditional farming.

The first step toward the revitalization of local communities should be the heritagization of symbolic localities and sites, the majority of which, however, were not preserved due to neglect. Part of the proposed heritage site falls under the authority of the current Lešť Military Training Area. Despite this, in the surviving, as well as the abandoned, localities, various elements can be found that could constitute a basis for the proposed musealization. Although incomplete, they still serve as concise and accurate testimonies about a vanishing culture.

Old cemeteries appear to be highly symbolic places, while the richly decorated and shaped tombstones may serve as a testimony to their times.

[3] This fieldwork has been conducted by Marta and Ján Botik (Botiková, 2016). Ethnographic findings from older research by Soňa Švecová are reflected in work of Ľubica Volanská and Juraj Majo (Volanská and Majo, 2016).

Fig. 2 The renovation of decorated wooden grave markers. (Photo: Ján Aláč)

In the 1870s, the inhabitants[4] began to decorate the graves of the deceased with richly decorated tombstones and wooden grave markers of distinctive shapes.[5] The altar-like tombstone design is typical of the 40-year period between 1870 and 1911, while the shaped and decorated wooden grave markers are characteristic of the following 20 years (Fig. 2). The graves from this period (1911–1930) provide testimony to the impact of the one-child practice in the locality: they are characterized by a greater proportion of young women's graves.

The project is based on the transformation of the burial site at an old cemetery in Horný Tisovník, and later also in Dolný Tisovník, both of which have gone unused for 100 years, into symbolic places. They should serve as examples of the skill, artistry, and esthetic sensibilities of their creators, a tribute to traditional stonecutting, but also the decline of

[4] Evangelicals of the Augsburg Confession (a Lutheran denomination).

[5] They only occurred in the following villages: Horný Tisovník, Dolný Tisovník, Červeňany, Madačka, Nedelište, Šula, Senné, Ábelová, Polichno, Praha, Veľký Lom, and Suché Brezovo, as well as Lešť and Turie Pole—villages that have since ceased to exist.

communities in Novohrad. This kind of in-situ heritagization element is an ideal way to commemorate the ancestors of the current inhabitants. With regard to the one-child practice, paradoxically although logically, these cemeteries testify to the dense population of the end of the nineteenth and beginning of the twentieth centuries, and the ways in which the area was subsequently depopulated (Aláč, 2007).

The renovations of the symbolic space included cleaning up the area, uncovering and stabilizing the tombstones, and complementing the space with replicas of wooden tombstones. These were followed by other acts of uncovering the memory of the land, as well as the fates of the local community members, in the form of placing additional elements typical of cemeteries in this place, as well as musealizing elements such as wooden fencing around the graves, symbolic glass balls installed onto graves, carved wooden boxes with accompanying texts, and information boards. The oldest burial site in Horný Tisovník, the so-called pandemic cemetery, where people stopped burying the deceased after the cholera outbreak in 1873, was also made accessible.

Revitalization of Symbols, Community, and the Cultural Landscape

The feedback, or response, to the implemented restoration projects, heritagization, and revitalization of the historical cemeteries and tombstones in Horný Tisovník and Dolný Tisovník was seen almost immediately. Only a year after the project had been implemented, flowers and lanterns began to be placed on the graves, which were previously inaccessible, and not only on All Hallows' Day. This illustrates the response from the family and relatives—descendants of the buried people. The direct positive response was also a result of involving the inhabitants in the project's implementation, which—given the small number of residents of the municipality—meant that a substantial part of the local population was involved, directly or indirectly (Aláč, 2011). The relatives and descendants thus found their way to the eternal resting place of the deceased.

The media continues to show great interest in the revitalization project, and the scientific community has also responded. The project was presented to landscape, environmental, and other related institutions.[6] Within the first two years of work (2009–2010), the character and implementation of the project were awarded with the 2010 Monument and Museum Award in the Restoration category. The informative value of the project in relation to the land was awarded the Special Recognition Award

Fig. 3 Involvement of the native population in a volunteer project in Horný Tisovník, 2017. (Photo: Ján Aláč)

[6] The project has earned the following accolades:

- The 2014 Slovak Republic Landscape Award (announced by the Ministry of Environment of the Slovak Republic, national coordinator of the award: the Slovak Environmental Agency) for the renewal of the historical memory of a locality; for rediscovering the historical memory, identity, and symbols of a locality in the countryside through the legacy of the past.
- 2017 Village of the Year, announced by the Ministry of Environment of the Slovak Republic/Slovak Environmental Agency/Village Renewal Association/Association of Slovak Towns and Villages.
- Special prize for rediscovering and renewing the traditional tombstones, and for developing the spiritual legacy of the past.

in 2014, and Horný Tisovník was awarded for the project Restoration of the Historical Memory of a Locality in recognition of its "rediscovering the historical memory, identity, and symbols of the locality through reading of a message from the past," and was renominated for the same prize in 2016. In the 2017 Village of the Year competition, Horný Tisovník was awarded a Special Prize for rediscovering and restoring the traditional tombstones (Fig. 3), as well as for enhancing the spiritual legacy of the past.

These awards and the publicity attracted by the project, which pointed out the distinctiveness of cultural elements independently or in their relation to the land as something exceptional, were passed on to the environment of Horný Tisovník and Dolný Tisovník. For the wider public, as well as the professional community, what was interesting was the contrast between the populated past and the depopulation of many settlements in Novohrad, which was difficult to understand at first glance. Questions as to why or where the population has gone, and the answer in the form of the one-child phenomenon typical of this region, make this issue and the place even more interesting, especially for those without any previous knowledge of the one-child practice. This cause remains known to a small circle of anthropologists, and a small number of the remaining population.

Environmental Communication and the Symbolic Language of Cemeteries

If we were to look for a symbolic place—one that embodies the memories that communities and landscapes themselves possess—we would be hard pressed to find places more apt than cemeteries. When we work with the concept of cultural landscape heritage (Aplin, 2007; Stoffle et al., 2016), we imagine preserved manifestations of mutually balanced interactions of nature and culture. In the symbols decorating the tombstones, we see a close pan-generational connection with the environment; the birth and death dates speak to the cruelty of time, and the texts are often a testimony for future generations.

As Robert Cox (2010, p. 20) notes,

> If we define ***environmental communication*** as simply 'talk,' or the transmission of information about the wide universe of environmental topics [...] our definition will be as varied as the topics for discussion. A clearer definition takes into account the distinctive roles of language, art, photographs, street protests, and even scientific reports as forms of symbolic action.

Cox proceeds to broadly elaborate the importance of symbolism in communication, referring to the Shannon–Weaver model of communication (Shannon & Weaver, 1949). The lack of understanding of the symbolism and function of symbolic places in a cultural landscape is related to concerns that an uninterested observer will not understand the meaning of the frame of reference of the place to which the symbolic message is directed. Local communities influence an individual's life cycle beyond the causal environment where the individual lived. Their responsibility and appeal to the wider community does not end with the biological death of individual actors in the dialogue between culture and nature. On the contrary, the talk of the dead is much more urgent. Their statements in the seminar bear a recognizable "anthropological trace" of successful or less successful adaptation strategies in space and time in the landscape.

As we can recognize in the human behavior in the restoration of abandoned cemeteries in depopulated areas, the primary anthropological function is no longer the survivors' care for the graves of their ancestors, but the dialogue of the survivors with the deserted land. The displaced or forgotten heritage made visible by the cultural landscape and the culture of local populations thus communicates vividly and enduringly. Out of the tension surrounding the mysteries of death, the idea has been born of a soul that exists, can return to the living, and do good or evil. Therefore, tombstones are covered with symbols representing good, protecting against evil, symbols of the eternity of the world (the Sun, the Moon), life and good deeds (trees, branches, flowers, fruits), souls deprived of a body, symbols of dying young (birds, beetles), love, and grief (a heart, a dead dove).

During the cemeteries' restoration, certain questions naturally arose: Who were these people? Where did they live, and why did they leave these places? Why are there so many deserted meadows and fields? The answers are found only in the symbolism of the tombstones. In poetic inscriptions, past generations encoded the loneliness and abandonment of their dead's final days, their concern for abandoned homes, and the desolate landscape. The tombstones and grave markers are particularly remarkable in terms of art—they are a collection of signs and symbols of the traditional farming environment. They have anthropomorphic and zoomorphic motifs such as birds and beetles, but there are also rich plant and geometric ornaments. There are engravings of highly stylized round flowers, but

Fig. 4 Epitaphs on the Horný Tisovník tombstones. (Photo: Ján Aláč)

also more realistic images, such as shrubs with flowers, fruit, and branches. Other common images include stars, solar motifs, and depictions of human figures and heads. The symbol of the clock face speaks of the relative relationship between the time of human existence and the time of death. Symbolic inscriptions are a celebration of the simple man, his life and faith, a celebration of nature incorporating moments of death and transience, as well as hope, for example, in motifs of rebirth. Their aesthetic stems from the want and simplicity of the material world in juxtaposition with the richness of the spiritual world.

Folk taxonomy is also used in a symbolic appeal to future generations to act responsibly. Epitaphs (Fig. 4) of the dead, revealing the results of the one-child practice, speak for themselves:

> "Here lies the flower that left the world abruptly and in haste."
> "In the spring of life, the morning of life, the beloved Rose withered and left."

> "Because they loved him and had him very much in their hearts, they adorned his grave with this rock, which stands here in that deed, in that temporality."
> "Like the withered flower of my youth at my most beautiful age, I lay my bones in the grave."
> (Turie Pole & Senohrad localities, 2019, translation Ján Aláč)

Contemplating the symbols of restored cemeteries inspires the imagination to reflect upon changes in the surrounding culture and country. The greater proportion of graves of young childless women and children corresponds to the gradual abandonment of the landscape. A syncretic idea of the effects of the loss of vitality of both culture and landscape is thus formed.

Discussion: Symbolic Dialogue in Environmental Communication

The collective memory of local communities is not formed by a mechanical sequencing of events, but by the enumeration of actions essential for lasting survival, a plan of symbolic links. Local arguments are subsequently based on such a perspective. In an effort to understand the complexity of the interconnection of communities with the landscape, the dialogical process of creating and understanding the completeness of meanings is revived, forgotten, and rediscovered yet again. We recognize this as Bakhtin's model of dialogic interactions (Bakhtin, 1990). The meanings of culture in the local context are created through the experienced multiplicity of its members' opinions, which are constantly being finalized and constantly coexisting in a continuous process. If this process disappears, for example, by depopulation as in this case study, the dialogue can continue, but only on a symbolic level, by transferring between new actors—heritage visitors. However, the knowledge contexts based on the experiences of these new participants' cultural heritage are limiting. Anthropological approaches to improving perceptions of the problem of rural depopulation can still draw on Bakhtin's original concept of getting to know the thing (landscape) and getting to know the inner world of the actors of present and past life. Therefore, to understand the relationship between culture and landscape, anthropologists need to communicate with native and local populations, enter into their worlds, and thus develop an active dialogical understanding.

Discussions about the demographic balance of the environment and the continuity of cultural values acquired through long-term adaptations of human communities—local and global—are reflected in the works of anthropologists and scholars of environmental communication. To understand the links between today's local communities and the changing environment, it is essential to clarify, in cooperation with scientists, the actions and behavior of local communities in times of scarcity, the motives for increased migration, the losses in social/ecological knowledge, as well as the perception of global changes. In field discussions, we—along with members of local communities—advocated for protection of historical landscape structures which serve as testimony due to the relation between their geomorphological conditions and farming techniques. To what extent, however, can the continuing depopulation of large areas and the change in landscape be resisted by turning to tradition?

From anthropological findings we can predicate that decline in local culture and landscape perception generally encourages symbolic thinking, while repeatedly referring back to the ideal or idealized world of the past. Through cultural, ecological, or environmental heritagization, individuals are capable of motivating their communities toward change and revitalization of seemingly lost values. These activities are noticed by more and more people from industrial and urbanized environments. We presume that these activities may help absorb a certain environmental contradiction felt by contemporary humanity. The question remains whether this distress is due to actual oversaturation of the physical or social space by psychological niches, or due to mental oversaturation, that is, regarding not our immediate experience, but rather our impersonal knowledge of global overpopulation.

Acknowledgments The authors gratefully acknowledge the contribution of the Slovak Research and Development Agency (APVV) within the project "The Social and Cultural Impact of Depopulation in Slovak Regions."

References

Aláč, J. (2007). Ľudový ornament na náhrobníkoch Novohradu. [The folk ornament on Novohrad tombstones]. *Pamiatky a múzeá, 7*(4), 24–27.

Aláč, J. (2011). Obnova starého cintorína v Hornom Tisovníku. [The reconstruction of the old cemetery in Horný Tisovník]. *Pamiatky a múzeá, 2011*(3), 29–31.

Andorka, R., Horská, P., & Head-König, A.-L. (1998). L'Europe centrale. Central Europe. In J.-P. Bardet & J. Dupaquier (Eds.), *Histoire des populations de l'Europe II. La révolution démographique 1750–1914* (pp. 427–461). Fayard.

Aplin, G. (2007). World heritage cultural landscapes. *International Journal of Heritage Studies, 13*(6), 427–446. https://doi.org/10.1080/13527250701570515

Appadurai, A. (1986). *The social life of things: Commodities in cultural perspective.* University Press.

Appadurai, A. (1996). *Modernity at large: Cultural dimensions of globalization.* University of Minnesota Press.

Bakhtin, M. (1990). *Bachtinskij sbornik.* [The Bakhtin Collection]. Moscow: Prometej.

Botiková, M. (2016). Reproductive behaviour in the first half of the 20th century in "one-child" regions. *Slovak Ethnology, 4*(64), 486–499.

Boyd, R., & Richerson, P. (2005). *The origin and evolution of cultures.* Oxford University Press.

Buday, D. (1909). *Az egyke Baranya varmegyeben [The system of single-child family in Baranya County].* Taizs.

Cox, R. (2010). *Environmental communication and the public sphere.* SAGE Publications, Inc..

Esping-Andersen, G., & Billari, F. (2015). Re-theorizing family demographics. *Population and Development Review, 2015*(41), 1–31.

Gajdoš, P. (2016). Developmental changes in Slovakia's socio-spatial situation. In V. Szirmai (Ed.), *"Artificial towns" in the 21st century: Social polarisation in the new town regions of East-Central Europe* (pp. 83–131). Institute for Sociology Hungarian Academy of Sciences.

Grulich, J., & Zeitlhofer, H. (1998). Migrace jihočeského obyvatelstva v období před třicetiletou válkou a po jejím ukončení. [The migration of the South Bohemian population in the period before the Thirty Years' War and after its end.]. *Historical demography, 22*(198), 79–106.

Hajnal, J. (1965). European marriage patterns in perspective. In D. Glass & E. Eversley (Eds.), *Population in history: Essays in historical demography* (pp. 101–143). Arnold.

Holec, R. (2003). Slovensko v ekonomike Uhorska [Slovakia in the Hungarian economy]. In E. Mannová (Ed.), *Krátke dejiny Slovenska [A concise history of Slovakia]* (pp. 221–227). Institute of History.

Ingold, T. (2000). *The perception of the environment: Essays on livelihood, dwelling and skill.* Routledge.

Kárníková, L. (1965). *Vývoj obyvatelstva v českých zemích 1754–1914.* [Population development in the Czech Lands, 1754–1914]. Prague: Nakladatelství Československé akademie věd.

Laslett, P., & Wall, R. (1972). *Household and family in past time*. Cambridge University Press.

Lawson, D., & Borgerhoff Mulder, M. (2016). The offspring quantity–quality trade-off and human fertility variation. *Philosophical Transactions of the Royal Society B, 371*(20150145). https://doi.org/10.1098/rstb.2015.0145

Lesthaeghe, R. (2010). The unfolding story of the second demographic transition. *Population and Development Review, 2010*(36), 211–251.

Livi-Bacci, M. (2000). *The population of Europe*. Basil Blackwell.

Massey, D. (1990). Social structure, household strategies, and the cumulative causation of migration. *Population Index, 56*(1), 3–26.

Moravčíková, D., & Fürjészová, T. (2018). Ecovillage as an alternative way of rural life: Evidence from Hungary and Slovakia. *European Countryside, 10*(4), 693–710.

Murin, I., & Kandert, J. (2018). The daily rhythm of the villager of Central Slovakia: Comparative study of the chronological ethnography of four families from the Sihla Plain in 1967–2017. *Český lid, 105*(1), 3–23.

Pavlík, Z. (2018). 1000 years of demographic development of Czechia and Slovakia. *Demografie, revue pro výzkum populačního vývoje, 2018*(60), 155–160.

Sear, R., Lawson, D., Kaplan, H., & Shen, M. (2016). Understanding variation in human fertility: What can we learn from evolutionary demography? *Philosophical Transactions of the Royal Society B, 371*(20150144). https://doi.org/10.1098/rstb.2015.0144

Shannon, C., & Weaver, W. (1949). *The mathematical theory of communication*. University of Illinois Press.

Špulerová, J., Drábová, M., & Lieskovský, J. (2016). Traditional agricultural landscape and their management in less favoured areas in Slovakia. *Ekológia/Ecology, 35*(1), 1–12. https://doi.org/10.1515/eko-2016-0001

Stoffle, R., Seowtewa, O., Kays, C., & Van Vlack, K. (2020). Sustainable heritage tourism: Native American preservation recommendations at Arches, Canyonlands, and Hovenweep National Parks. *Sustainability, 12*(23), 9846.

Stoffle, R. W., Arnold, R., & Bulletts, A. (2016). Talking with nature: Southern Paiute epistemology and the double hermeneutic with a living planet. In G. Tully & R. Mal (Eds.), *Collaborative heritage managements* (pp. 75–99). Gorgias Press.

The Permanent European Conference for the Study of the Rural Landscape. (2020). http://www.pecsrl.org/index.html

Ustaoglu, E., & Collier, M. (2018). Farmland abandonment in Europe: An overview of drivers, consequences, and assessment of the sustainability implications. *Environmental Reviews, 26*(4), 396–416.

Vasary, I. (1989). The sin of Transdanubia: The one-child system in rural Hungary. *Continuity and Change, 4*(3), 434–464.

Voľanská, Ľ., & Majo, J. (2016). The one-child pest is now affecting the body of the Slovak nation: On restricted reproduction in southern Slovakia in the first half of the 20th century. *Ethnologia Slovaca et Slavica, 37*, 41–59.

Westhoek, H., van den Berg, M., & Bakkes, J. (2006). Scenario development to explore the future of Europe's rural areas. *Agriculture Ecosystems & Environment, 114*(1), 7–20. https://doi.org/10.1016/j.agee.2005.11.005

Zeitlhofer, H. (2006). Migrace na českém venkově na počátku 20. století. [Migration in the Czech countryside at the beginning of the 20th century]. Supplement. In E. Maur & J. Grulich (Eds.), *Dějiny migrací v českých zemích v novověku* [The history of migration in the Czech Lands in modern times]. *Historical Demography*, 123–144.

Living Stone Bridges: Epistemological Divides in Heritage Environmental Communication

Richard Stoffle

This chapter presents the cultural perceptions of five Native American tribes and pueblos who communicated their environmental understandings of natural stone bridges during an ethnographic study funded by the National Park Service (NPS). Natural Bridges National Monument (the park) is located in southern Utah, USA. The participating tribes and pueblos stipulate that this is an ancient World-balancing location dominated by Bears Ears Buttes on the flank of the Abajo Massif, the two rivers who carved the bridges, and the massive living stone bridges themselves (Fig. 1). World-balancing ceremony areas have special significance to contemporary Native Americans. Spiritual leaders traveled long distances to such areas when extremes in weather, ecological health, and human social

---

R. Stoffle (✉)
Bureau of Applied Research in Anthropology, School of Anthropology, University of Arizona, Tucson, AZ, USA
e-mail: rstoffle@arizona.edu

A. Sjölander-Lindqvist et al. (eds.), *Anthropological Perspectives on Environmental Communication*, Palgrave Studies in Anthropology of Sustainability, https://doi.org/10.1007/978-3-030-78040-1_7

Fig. 1 Owachomo Bridge, with a 180 ft. (65 m) span, 106 ft. (32 m) tall, 8 ft. (3 m) thick, and 27 ft. (8 m) wide at the top, and alive since Creation. (Photo: Richard Stoffle)

relationships called for the world to be set back into balance. This area and its associated plants, animals, water, minerals, rock peckings, rock paintings, artifacts, old kivas, and special topographic features define how this heritage cultural landscape was Created to maintain the Earth. As such, these resources and the place where they occur are a unique heritage foundation for contemporary Indian people. The participating tribes and pueblos are willing to engage with the NPS to recommend culturally appropriate interpretations and ways to protect this heritage park because of the cultural centrality of this area.

Heritage Environmental Communication

The study contributes to a 140-year-old anthropological tradition that involves talking with people about themselves and their aboriginal environment adaptations and subsequently conveying these understandings to professional and general audiences. Cross-cultural communication of deep culture (Lynch, 1996) that derives from *thick description* (Geertz, 1973) is difficult, and as such resembles the *Deep Ecology* philosophy established by Norwegian philosopher Arne Ness (Ness, 1973) and Sessions (1987). Like the current Deep Ecology Movement (Drengson & Inoue, 1995), Native Americans tend to view the Earth as a single whole whose parts are integrated, and which has human-like rights. Critics of the Deep Ecology Movement and indigenous cultural views, such as those presented here, have argued that the observations are not scientific findings but instead are religious interpretations, and thus are not useful knowledge (Keller, 2009). Despite critics of these types of philosophies, today there are graduate programs, such as one at the University of Oslo in the Center for Development and the Environment, that build upon the deep ecology movement (University of Oslo, 2020). Similarly, there are dozens of graduate programs, such as American Indian Studies at the University of Arizona (UofA), that build on Indigenous traditional cultural knowledge. Environmental communication across the epistemological divide created by these views is a foundation of this chapter.

Our analysis further draws on that of Clifford Geertz, who provided the social sciences with an understanding and appreciation of thick description (1973). Geertz applied thick description to anthropological studies, particularly to his own *interpretive anthropology* perspective. He urged anthropologists to consider the limitations placed upon them by their own cultural cosmologies when attempting to offer insight into the cultures of other people. He produced theory based on the notion that culture is essentially semiotic, that is, composed of signs and symbols, and it requires more than just words to convey the intended and imbedded meanings and deep understandings (Geertz, 1973). Thick description is required to interpret cultures.

Ecological communication is a foundation of cultural anthropology. For example, after 1881 Franz Boaz embraced *anthropogeography* as an holistic field of study; a perspective that he applied in North Africa, western Asia, and Baffin Island. He continued to use this analytical frame as a supervisor of the Jesup North Pacific Expedition to understand the peoples of Siberia and Canada for the American Museum of Natural History from 1897 to 1903 (Lowie, 1937, pp. 128–136).

Bronislaw Malinowski used systematic interviews and participant observation beginning in 1914 to understand the traditional people of the Trobriand Islands. His work was holistic in the sense that any aspect of life was to be understood itself, as well as its relationship to other aspects of the environment where people lived. He maintained that anthropologists must have the goal of grasping the ***native point of view***, understanding their relationship to life, and realizing their vision of the world (Malinowski, 1922). He observed that fishing in less dangerous lagoons required fewer religious restrictions than was required for fishing in the dangerous open oceans, which was an early observation of cultural adaptations to environmental differences (Malinowski et al., 1935). He would grapple with the applications of ethnographic findings to understanding and explaining how cultures change (Malinowski & Kayberry, 1946). A century ago, Malinowski's research documented his basic commitment to understanding peoples' relationships with their environments and effectively communicating these to the profession and public.

Ethnography is a holistic description of a society and so it has always had an environmental component. Lowie (1937) defined ethnography as the science that studies "the sum total of what an individual acquires from his society." It depicts "the whole of cultural reality," like a natural science.

George Murdock, from the 1920s until the 1940s, suggested that ethnographers ask a set of similar questions so their studies could be compared. His arguments for systematic ethnographic field work were presented in his *Outline of Cultural Materials* and later a more specific guide called the *Survey of Cultural Elements* (1920–1940). The latter contained over 4400 ethnographic questions; including an entire section focused on the environment (Murdock et al., 1982). Murdock helped established the Human Relations Area File archive at Yale University, where all ethnographic reports were to be archived, divided by topics to permit easy comparisons of cultural patterns across many cultures. He began the journal *Ethnology* in 1962 for publishing comparative findings which were seen as the building blocks of anthropology as a science. There have always been human ecology and cross-cultural communication components in cultural anthropology.

The findings of this study are situated in the conceptual notion of ***heritage environmental communication***, which is understood as a type of environmental communication that often occurs during contemporary cross-cultural communications when traditional people or scholars explain ancient relationships with the environment as these were established over hundreds or even thousands of years (Tilley, 2010; Yearly, 2006). A New Zealand study which assessed the role of cross-cultural communication in collaborative partnerships with the Maori people found that it is essential to recognize the basic premises of Maori cultural guardianship and traditional knowledge (Lyver, 2005).

Interpreting past environmental knowledge and use patterns is a different challenge for environmental communication. Tilley (2004), for example, explains how and why thousands of years ago Europeans interacted with massive stones called *menhirs*. Cummins, Farmer, and Russell (2013) bring forward in time cultural understandings across hundreds of years in Barbados, West Indies. Antoinette (2012) reaches back in time to *speak for the enslaved* and their relationships with plantation environments along the Atlantic coast of the USA.

Native Americans in this chapter recount songs, stories, and natural relationships of their ancestors based on contemporary oral history and simply listening to voices recorded in stones. Humans develop social constructions of nature which imbue it with cultural meanings, establish appropriate patterns of interactions, and argue for preservation goals. These social constructions are the foundation for selected aspects of nature becoming key components in the heritage of a people.

In this chapter, environmental communication about natural heritage places and resources is considered different from communication about anthropogenic places where either there are special archeological residues or the place itself has been created as an artifact. Debates over the meanings of places or connections to them by contemporary peoples often can be resolved by using artifacts as evidence. Heritage communication about natural resources is different, especially where there are no associated artifacts and thus it is not possible to prove why they are culturally significant to people. Instead, traditional cultural logic and oral history must be used to establish what Tilley (1994) calls the *phenomenology of landscapes*. This is important because questions of association and meaning of non-artifactual natural places and resources are either believed or not. These officially are called Traditional Cultural Properties in the USA.

Others have documented the heritage value of natural resources that are not associated with artifacts. These include flying fish in Barbados, West Indies (Cumberbatch, 2013); sacred trees, mountains, and water sources in Nigeria (Borokini, 2016); water resources in Bahrain (Rudolff & al Zekri, 2014); falcons in Mongolia (Soma, 2012); and Paektu, a volcanic mountain on the China–North Korea border, which is the spiritual home for Korean people (Winstanley-Chesters & Ten, 2016). See also Yoko Kugo's analysis of indigenous places, their meanings, and names in Alaska (chapter "Community Voices, Practices, and Memories in

Environmental Communication: Iliamna Lake Yup'ik Place Names, Alaska" of this book), where she discusses why it is important in heritage environmental communication to use native names and landscape understandings.

For this chapter, we consider three factors that make heritage environmental communication difficult (Stoffle et al., 2004). First, some problems occur because actors differ in terms of their knowledge of the issue. Resolution in these cases can occur through education, usually the scientists educating the lay persons. Second, other problems occur when people accurately perceive and agree on what is out there and value similar natural features but rank some above others and thus have different outcome goals. Here the *Nature Services* debate is instructive (Aisher & Damodaran, 2016; Holzman, 2012; Peterson, 2012). All agree that the components of nature have various positive benefits for other natural features as well as for humans. One perspective, however, prioritizes Human Services over Nature Services. Biologists tend to take the reverse position. A third type of problem occurs because the actors have different culturally based beliefs regarding what resources are involved; that is, what even exists in nature—how it is formed, whether or not it is sentient, and how it contributes to human heritage (Goldman, 1999; Jackson, 1981). When these cross-cultural social constructions of natural reality are fundamentally different, we can have an epistemologically derived problem (Stoffle et al., 1990).

There is a fourth factor influencing heritage environmental communication, which is merely mentioned here because it is beyond the scope of this chapter. Joosse and her fellow scholars (2020, p. 6) call this *Discursive Colonization*, that is, the reproduction of the interests of the powerful through certain narrowly defined forms of knowledge and scholarship. Sometimes heritage communication does not get heard because it is politically suppressed. A postcolonial world emerged after the 1960s and expanded agency for traditional peoples in the early twenty-first century (Antoinette, 2012; Cummins et al., 2013). New voices have emerged to decolonize traditional lands (Smith, 2012) and bring what Aja Martinez (2020) terms the *Counterstory*.

This chapter is focused on how the notion that *the Earth is alive* can become a fundamental barrier to heritage environmental communication. This is an epistemological foundation of Native American beliefs and culture, or what Rappaport (1999, pp. 263–71, 446) calls an *ultimate sacred postulate* and what Goldman (1999) calls a philosophical *primitive*. The concept of a living universe is essential for understanding native culture

(Stoffle et al., 2016; Stoffle & Zedeño, 2002, p. 174). The universe is alive in the same way that humans are alive and fully sentient. It has physically discrete components that we call *elements*, and an energy source that brings them alive that is called *Puha* in the Numic language, something that can be translated as "Creation energy" or "power" (Stoffle & Zedeño, 2002). Elements like mountains, rivers, fauna, flora, wind, and minerals have most of the same characteristics as humans, including the ability to communicate, to help other elements, the power to accomplish their own goals or *agency*, and even the capacity to lie. The stone bridges in the park are understood by Native American people as *living beings* through this epistemological perspective.

The Case Study

Native Americans have stipulated that there are *living stone bridges* in Natural Bridges National Monument, Utah, USA (Fig. 2), and in doing so they thus have framed a heritage environmental communication

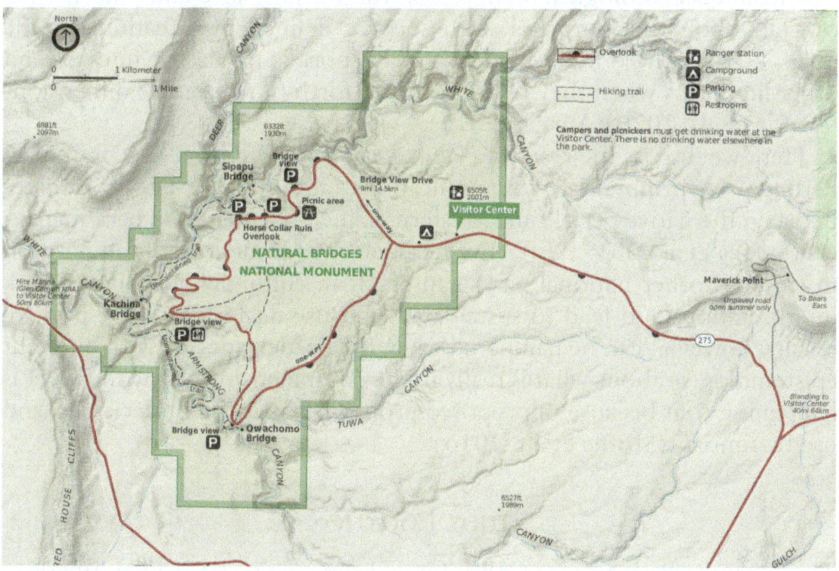

Fig. 2 Map of Natural Bridges National Monument. (This public image used with permission of the National Park Service)

challenge based on epistemological differences between themselves, park managers, and Western scientists. The Congress of the United States in 1908 identified three massive stone bridges as charismatic features of the natural landscape, and they thus qualified as a national heritage place worthy of federal protection by the NPS. This decision was based on the premise that these stone bridges are natural products, having been made by rivers and weather, and should be protected in a ***nature park***. The stone bridges are managed with the advice of geologists and interpreted by the park managers as spectacular natural features.

In 2018–2020 the NPS funded an Ethnographic Overview and Assessment (EOA) involving representatives from a number of tribes and pueblos in order to officially have Native American representatives share their heritage cultural perceptions about park plants, animals, water, archeology, and the three stone bridges. A total of 292 ethnographic knowledge sharing events documented that the placement of the stone bridges on two rivers and at the foot of a mountain was a plan of Creation, and that each bridge is alive and sentient with its own agency and personality. These heritage cultural stipulations are in keeping with a common Native American epistemological premise that the world is alive and reflects (i.e., can talk about) its purposes as these were defined by Creation (Stoffle et al., 2016). The bridges further serve as portals for use by humans and other life forms when they travel back and forth to alternative dimensions (Stoffle et al., 2020a; Varner, 2012).

Here we discuss the resulting ***epistemological divide*** that now frames all heritage environmental communications about these stone bridges between the native people and Western-trained scientists and park managers (Stoffle et al., 2017). Effective heritage environmental communication in this and other national parks can result in culturally appropriate park interpretations and lead to sustainable land-use decisions. These positive results, however, can be hampered when the participants act with different epistemologies about what is really in the natural environment and what is happening to it because of proposed projects, management decisions, and use by tourists (Stoffle et al., 2016).

Study Location

The National Park Service (NPS) funded the park's first Native American EOA study in 2018–2021 (Stoffle et al., 2020b). U.S. President Roosevelt in 1908 defined the park's meaning when he proclaimed:

> WHEREAS, a number of natural bridges situated in southeastern Utah, having heights more lofty and spans far greater than any heretofore known to exist, are of the greatest scientific interest, and it appears that the public interests would be promoted by reserving these extraordinary examples of stream erosion with as much land as may be necessary for the proper protection thereof.

This establishment legislation by Congress that identified the park as a geological wonder continues today. This is the Western science-based foundation theme presented in the park's interpretative video for tourists, adult and youth reading materials sold in the Visitor Center, and interpretive displays located along park walking trails and overlooks. According to this theme the park is a place of unique and naturally formed massive stone bridges carved by running water.

Southern Utah topography is dominated by high uplifted volcanic massifs, broken sedimentary plateaus, and deeply gouged canyons. The Colorado River flows from the Rocky Mountains in the east to combine with the Green River from the north at Canyonlands National Park before they together flow through the Grand Canyon to the Sea of Cortez in Mexico. The park is located around two much smaller rivers, the White River and the Armstrong River, whose headwaters are generally located in the Abajo Massif, a largely snow-capped Sky Island standing high above the surrounding region. Bears Ears Buttes are a specific headwater for these rivers (Fig. 3).

The White and Armstrong Rivers have carved out a broken sedimentary plateau, making deep canyons before flowing into the Colorado River about 30 miles to the west of the park. Over eons these meandering rivers cut deeply into the soft sandstones. The rivers formed oxbow meanders which were eventually undercut, thus forming the three stone bridges: the Owachomo, the Kachina, and the Sipapu. These are the spectacular centers of the park (Fig. 4).

Study Methods

The NPS funded the EOA to understand the cultural meanings and importance of the natural and archeological resources within the park that are associated with Native Americans. The EOA was officially designed to be participatory (Joosse et al., 2020, pp. 6–8) so the tribal and pueblo representatives would share their opinions as to where the study visits

Fig. 3 Bears Ears Buttes from Park Mesa. (Photo: Richard Stoffle)

should occur and what were the most salient topics to be discussed. The study was also funded to meet certain park management and interpretation goals including (1) learning about tribal perceptions of the stone bridges, (2) knowing oral histories of the ancient farming homesteads in the center of the park, (3) understanding the intended purposes of the peckings and paintings near the bridges, and (4) hearing about the contemporary uses of the abundant and rare plants located throughout the park. Together these cultural perspectives potentially serve to inform park interpretations in the visitor center museum and along hiking trails and provide new ideas for park management, especially how tourists visitors should interact and treat heritage places like the stone bridges (Stoffle et al., 2020c).

This analysis is based on 292 ethnographic interviews with official representatives of four tribes and pueblos. Representatives were sent to participate in the study from (1) the Pueblo of Zuni, (2) the San Juan Southern Paiute Tribe, (3) Acoma Pueblo, and (4) the Ute Mountain Ute

Fig. 4 Kachina Bridge at confluence of White and Armstrong Rivers with rock panel and structures behind, group Kiva in bottom, and residential mesa on horizon. (Photo: Richard Stoffle)

Tribe, especially the White Mesa people. Data sharing occurred at six formal locations and wherever tribal or pueblo representatives wanted to stop and talk. These locations reflect the kinds of resources in the park, including overlooks into the canyon, a previously excavated traditional farm house on the mesa, a large painting of a Red Bear on a rock wall, abundant plant communities, spectacular landscape views including ones of the nearby Bears Ears Buttes, and the Visitor Center with a park movie and museum interpretations. Most interviews were taped with the permission of the representative to ensure accuracy. All tribal governments and representatives understood that no confidential information was being sought and that findings, after being reviewed by the tribe, would become public through new interpretative displays, the training of park rangers, and perhaps in an updated visitor movie.

Study Findings

Cultural perspectives shared by representatives describe the stone bridges as being situated in a matrix of nearby and functionally interdependent places. For example, the Red Bear pictograph panel, which is located on a sheltered face of a massive sandstone cliff, is high above the confluence of two rivers (Fig. 5). It has a panoramic viewscape that includes Bears Ears Buttes, which are a source of river water and a spiritual place on the Abajo Massif. The panel close-up (Fig. 6) clearly identifies small red dots that occur above the image. These dots also occur above the other three bear (black, yellow, and white bears located in the region outside the park) that are pecked and painted on similar panels suggesting their connection with star and planet alignments ceremonies.

Each related cultural place, the resources it contains, and the views between places add to the meaning of the stone bridges and an understanding of why the area has always been special to Native Americans. It is important to recognize that the area has been occupied for more than 13,000 years, as indicated by the nearby discovery of Clovis Period spear points (Pitblado, 1998). This is the time-depth frame within which most

Fig. 5 Red Bear painting panel high above confluence of rivers. With viewscape of Bears Ears Buttes. (Photo: Richard Stoffle)

Fig. 6 Close-Up of Red Bear painting panel. (This image used with permission of Joseph Kayne Photography (www.josephkaynephoto.com). © Joseph Kayne)

oral history events are referenced, but there is a time before time that is also relevant. Both times can be accessed by properly prepared and trained spiritual people through a stone bridge portal.

The cultural perspectives of participating tribes and pueblos represent a range of ideas about the park; however, there was general agreement regarding many issues. One of these is the interpretation of the bridges as being alive. The bridges are surrounded by other living elements like Bears Ears Buttes, the rivers and springs, minerals, medicine plants, and spiritual animals, all of whom were made and given purpose at Creation. Native Americans who came to this area in the past and their descendants today recognize that the matrix of significant elements was placed here at Creation to attract humans to the area. The energy of all resources stimulates their interaction with each other and brings humans to conduct ceremony. The resources participate in various kinds of ceremonies, but clearly this is a place designed for the conduct of world-balancing ceremonies.

The following are EOA study statements from participating tribal and pueblo representatives. These have been selected for this analysis but have not been edited.

Zuni

- Areas such as Owachomo Bridge are a part of the Zuni cultural landscape, as the area itself is home to many shrines and offerings that further cement the link between the community and the monument.
- Kachina Bridge is a highly ceremonial place. This is indicated by the confluence of Armstrong River and White River, the great kiva depression nearby, and the painted walls of the small structure by the bridge. Zuni people still place certain images on walls of highly ceremonial structures today.

Southern Paiute

- The natural bridges represent one of the origin places of the Southern Paiute people, which some believe took place at Rainbow Bridge, located southeast in the Grand Canyon. This was the location where the different people of today were all one people with one language, before the split off and the different languages of the world were created. Songs, prayers, and offerings which are Southern Paiute practices associated with natural bridges were often performed by religious leaders.
- The ethereal characteristics of these bridges are so powerful that the bridges occupy an immense presence within the community. As a sign of respect and awe certain tribal members will not pass under bridges in response to the power they hold. So are the social norms of the San Juan Paiutes. Going under or through them is against our teachings, unless you have a reason to be with the bridges. These bridges have a powerful role in the beliefs of the San Juan Paiutes which illustrates the years of interaction the community has had with these geological features.
- Rock peckings and paintings in the monument are not art. They demonstrate the narratives of the past inhabitants of this region. The Southern Paiute people made these markings when all of the tribes were one people, and they have continued this practice throughout time.

Acoma Pueblo

- Natural Bridges is one of the four stops made after the Acoma people's emergence from *Shipap* on their migration to present-day Acoma.
- The Acoma people's connection to these traditional lands is paramount for the overall health of the local and world ecosystems. Acoma people maintain that Mother Earth provides these resources through a reciprocal relationship with the Acoma people. Thus, anything that keeps us away from this area poses risks to both nature and Acoma identity.
- Sipapu Bridge, named after the Puebloan place of emergence—known as *Shipap* to the Acoma people, is a physical representation of their origin spot from the underworld, where the first people emerged into the current world. Acoma representatives agreed that the name is appropriate in this context.
- As large, geological features in this desert landscape, the natural bridges likely contributed to calendar-type observations, which are essential to Acoma lifeways, as they dictate farming and ceremonial practices. These activities would be performed by religious leaders.

Ute Mountain Ute

- Bridges are powerful altars. Ute people pray to natural bridges to bring strength to the area and use them in reciprocal transactions to bring blessings to themselves.
- Creation stories, such as the one of Bear, illustrate the Ute people's direct connection to and creation within this cultural landscape. We identify with archeology, including the Red Bear panel, and viewscapes, such as the Bears Ears Buttes (see Fig. 5). These features demonstrate the long-standing connection the Ute people had with Natural Bridges, as well as the significance of this sacred cultural landscape.

Analysis

Native American cultural perspectives are filled with stipulations that contrast with those held by Western science, especially geology and archeology, and thus are at odds with current NPS interpretations of this park, its history, and what is out there. The most obvious of these conflicting stipulations is that the park and its resources are alive and have been so since Creation. From the Indian perspective these park resources are alive without the presence of humans, and so are not social constructions. The park resources did not come alive because Indian people developed a social construction of the living universe, but instead the universe has been alive since Creation and Native Americans were taught this by the elements themselves and the Creator.

This premise contributes to an epistemological divide when it comes to learning more about park resources. Scientists take samples and study them in a laboratory to determine age and origin of stones, soils, and artifacts. Native Americans, on the other hand, sit for long periods near a resource and it tells them who it is and what it desires from park management and interpretation. The resource asks why these people are here and expresses concern or pleasure regarding their behaviors. It is the home of the resource and it thus has a right to direct visitor behavior.

The notion of culturally appropriate communication with park resources is very much contested. Native Americans during this study were attracted to a resources or place like below a stone bridge. Here they sat quietly for long periods to listen to the bridge recount its memories about past visits by Indian people and what they sang and said to it in ceremonies. The bridge is like an elder, who once she/he begins to talk is not interrupted. The bridge stories may take a long time for it to tell and the listener should be respectful. Elders expect respectful visitors to introduce themselves, explain why they are present, and share ideas. The act of heritage learning from elders involves cultural protocols and requires time and patience.

In one instance an elder lady from the Ute Mountain Ute tribe had just completed a study discussion while sitting below Owachomo Bridge. A German couple approached and asked the elder a question. She wondered if they would stay for the answer and they said yes. So the elder took more than 45 minutes to explain her people's relationship with Owachomo Bridge. After returning to Germany the couple sent a response regarding the exchange under the bridge describing it as the most important event during their months of visiting U.S. national parks (Stoffle et al., 2020b). This underlies the ongoing issue of what it can mean to learn about Native American culture in national parks and perhaps how it may be best for this kind of heritage communication to occur.

Culturally more accurate interpretations of park resources and history are key for achieving the dual purpose of educating the public and gaining their support for preserving parks. The EOA had a specific goal of documenting Native American assessments of what is currently said about cultural resources in the park so these potentially can be incorporated into new park films, brochures, ranger training, and interpretative displays.

Tribal and pueblo representatives expressed a desire to bring their voices into the park and so they shared hundreds of ideas for making park interpretations speak to Indian issues. This occurred in 37 taped idea-sharing sessions in the Visitor Center and elsewhere during hikes. One entire chapter of the EOA report (Stoffle et al., 2020b) is devoted to these recommendations; a few are presented here.

Native Americans recommended the following interpretation changes: (1) we never left this area we just live somewhere else today; (2) the area was sacred to many Native American cultural groups and the park should say this; (3) the bridges continue to be culturally important to us so we send them prayers and they send wisdom back to us; (4) the stone bridges are alive and can talk as can the plants, animals, and places where we lived in the ancient past so these should be treated with respect and guided by protocols jointly produced by NPS and Native Americans; (5) the park is not an accident of geology but is instead part of a larger plan to have a place for individual healing and world-balancing ceremonies.

These heritage interpretations are difficult for the park to discuss in displays, films, and ranger tours. The first three involve cultural affiliation, perceived continued sacredness, and the movement of prayers to and from the park. Spirituality is a difficult issue for the park to address because of the congressionally mandated legal separation of government and religion institutions in the USA and because there are few agency guidelines for dealing with such issues. The latter two conversations are even more difficult for the park because they argue against the Western science interpretation of the stone bridges as having been produced by river erosion. They also are troubling for the park because if the stone bridges are alive and sentient, they potentially become *someone* to engage with during land management decisions. This would involve bringing Native Americans to talk with the bridges and potentially incorporating the recommendations from the bridges into management (Stoffle et al., 2016).

Native Americans recommended management changes regarding the bridges, including (1) other people such as tourists can be in the park, but they should be told by the NPS that this is a special Creation place, and (2) before visitors walk under a stone bridge, they should stop, introduce themselves, and ask for permission to visit. The park will have major concerns about implementing these recommendations inasmuch as they involve both the recognition of a cultural view of Creation and spirituality and establishing public guidelines for interacting with stone bridges.

Discussion

The notion of an *epistemological divide* has been used here to explain why oppositional heritage environmental communication is fundamentally different from other kinds of communication problems. These are pervasive in Native American interactions with federal land managers, and thus constitute what is called *macro-evidence* by Gingrich (2009, p. 179), inasmuch as similar patterns have been documented in so many instances among so many kinds of native peoples.

After arguing for much of his career as a Native American spokesperson and lawyer, Vine Deloria (1997) wrote a polemic called *Red Earth, White Lies,* in which he describes a number of "extreme" cases where Indian people maintain some truth about the world that is disputed by Western science. In the book, Deloria stipulates that everything Indian people say about the Earth is *veritas* (the truth), which is what a lawyer does when beginning an argument.

Resolving these types of problems tends not to be a part of either the training or background of Native Americans, NPS personnel, or Western scientists. While anthropologists and other social scientists work on the complexity of cross-cultural communication (Gudykunst, 2003; Hall, 1959; Maffie, 2000, 2020), there are few research-based solutions for improving heritage environmental communications. Carroll (2014) suggests that Tribal Parks made and managed by native people offer alternative perspectives to conservation and communication by accounting for their own land-based epistemologies and practices. When the tribes control the interpretation and management of a park, they can directly explain and use their culture.

The Stone Bridges case illustrates that fundamental problems can derive from failures to communicate about and listen to culturally based differences in environmental perception. The case also raises the question as to whether or not more talking between cultural groups can bring understandings that are sufficient to (re)solve such problems. Geertz (1973, pp. 3–33) argues that the capacity to produce thick descriptions is needed for interpretation and communication of culture. Brody (1998, pp. 6–7) demonstrated that even when the anthropologist and the native expert speak the same language it is necessary to interpret colloquial speech. In their *Talking with Nature* analysis, Stoffle, Arnold, and Bulletts (2016, p. 94) concluded that Native Americans prefer that their knowledge be given equal weight as veristic reality along with (or parallel to) contrastive ones from Western science, but for the moment most Native Americans simply want to exercise significant control over the preservation and use of their traditional heritage natural resources and would like their stories to be accurately interpreted.

Lessons from other parks suggest a way forward when an impasse is reached on important communication issues. Traditionally the NPS relied on science as a foundation and test of the truth about park resources and culture history. Today, however, most national parks have shifted from finding one truth to presenting multiple perspectives on a question even if these disagree, which is called a *multi-vocalic perspective*. This communication adaptation is political in that it is a management decision to be inclusive and permit the public to recognize that there often is no single common resource interpretation or best management practice.

Each national park answers visitor questions about past use by contemporary Indian peoples, but typically contrastive views are provided by archeologists, historians, and certainly the Native American groups themselves. An EOA conducted at Sunset Crater National Monument near Flagstaff, Arizona identified a distinctive disagreement regarding a volcanic eruption. Professional archeologists and geologists describe the Indian response to the AD 1066 eruption as fearful and them as running away. A painting of running people with hands in air in fear was for decades prominently displayed in the visitor center. Indian people in the EOA study had the reverse interpretation. Volcanic craters are common in the southwestern USA, and are often called *earth navels* by pueblo people and responded to by most Indian people as places where the Earth is reborn (see Van Vlack, chapter "Dancing with Lava: Indigenous Interactions with an Active Volcano in Arizona" of this book). The Sunset Volcano was an active eruption that lasted for about 100 years. It became, according to EOA native representatives, a central place for spiritual leaders, who constructed areas for living, viewing, observing, and talking with the volcano in a new village built nearby at what is now Wapakti National Monument.

Sunset Crater National Monument resolved their communication disagreements by retaining the original displays with the Western science interpretation, but adding a heritage kiosk surrounded by seats. On the kiosk is a map of contemporary Indian groups who have expressed a connection with the park, and next to each of their names is a phone receiver connected to a taped message so the tourist can hear their view of the volcano in the native voice (Toupal & Stoffle, 2004). Such an approach is mirrored in other parks that provide multi-vocalic brochures, museum displays, and complex visitor films (Ruppert & Smythe, 2017).

The present analysis should be contextualized as both a part of an ongoing process of the NPS seeking culturally sensitive Native American interpretations of park histories and to understand complex native relationships with non-artifactual places. For example, Natural Bridges are similar to a massive single stone bridge called Rainbow Bridge, which is located about 125 miles to the southwest in a side canyon of the Colorado River (Jett, 1992). It was placed on the National Register of Historic Places (NRHP) in 1910 and became a National Monument to protect and celebrate the bridge as a natural wonder.

Although culturally associated Native Americans participated in Rainbow Bridge's "discovery" by Euroamericans, they were neither highlighted in the monument's interpretations nor involved in its management. More than a 100 years after the monument was established, the NPS hired anthropologist David Ruppert (2017) to conduct EOA interviews with five tribes and pueblos. The research produced findings similar to those of the Natural Bridges EOA. Subsequently Rainbow Bridge was placed on the NRHP as a Traditional Cultural Property and is now interpreted as both a natural wonder and a culturally central native place.

While political decisions have resolved some heritage communication disagreements, the management of national park resources is based on knowledge domains that are dominated by Western science. These university-trained subject experts advise park managers on when to do a controlled burn, how to stabilize an eroding archeology site, which beetle is killing the pine trees, what is causing declines in the rabbit population, and how to protect raptor nesting. Science specialists tend to doubt Native American recommendations about how to protect heritage natural resources. While there may be clear physical cause-and-effect interpretations of a given problem, from a Native American perspective the problem is often that the area is out of spiritual balance and the resources miss hearing ceremonial performances. Balancing resources, places, and even the world is, after all, why the Creator made the park and taught Indian people the appropriate ceremonies. Scientists also tend to doubt that Indian representatives can learn actionable environmental information by sitting with the resources and listening to their knowledge. Potential common ground for resolving these park management challenges can derive from Western science, national parks, and Native American understandings being either combined or used side by side.

An essential step for resolving heritage environmental communication barriers is to define their source and find common ground, even if it may be to agree to disagree or not to seek a single final answer. Easy solutions are unlikely to occur in cross-cultural communication involving epistemological differences, and so systematic ethnography and other social science research is suggested. The NPS and Western scientists cannot just sit down and talk with native representatives. Words are symbols according to Geertz (1973), and so to just hear a word is not necessarily to understand its full meaning. Cultural knowledge involves specific kinds of places, food, music, group discussions, and translation from one language to another, according to Hopi elder Emory Sekaquaptewa (personal communication). As such, native cultural knowledge is not unlike that surrounding Passover (also called Pesach) for Jewish people. New place-based research is required to identify and culturally explain during cross-cultural communication. The research should produce thick descriptions of environmental issues based on interpretations provided by Native Americans. By funding this EOA study, Natural Bridges National Monument and the National Park Service have expressed a willingness to listen to native perspectives and to consider new public interpretations and policies based on accurate cross-cultural interpretations.

Acknowledgments The Natural Bridges National Monument (Natural Bridges) Ethnographic Overview and Assessment (EOA) was completed with the assistance of many people and agencies.

The participation of the Pueblo of Zuni, the San Juan Paiute Indian Tribe, Acoma Pueblo, and Ute Mountain Ute Tribe made this study possible. The American Indian representatives who participated in site visits to study areas endured difficult hikes and long days, and shared knowledge about their cultural and natural resources. Thanks to the tribal and pueblo administrative officials who facilitated the study.

The study was conducted under the supervision of Jeanine McElveen and Vernon Cody, Park Superintendents, Southeast Utah Group, National Park Service, Moab, Utah. NPS subject experts Terry Fisk, Kelly Adams, Thann Baker, Laura Martin, Noreen Fritz, and other NPS employees participated. Canyonlands Natural History Association volunteers contributed in many ways.

This chapter was prepared at the Bureau of Applied Research in Anthropology (BARA) in the School of Anthropology at the UofA. Thanks to the administrative staff in the business office who facilitated NABR budgets, travel advances, payments for independent contractors, and other financial details. Office staff included Cynthia Malbrough, Laura de Steiguer, Christine Duddleston, Juliana Reddick, and Nick Smith. School of Anthropology Professor and Director Professor Diane Austin facilitated the project.

References

Aisher, A., & Damodaran, V. (2016). Introduction to human-nature interactions through a multispecies lens. *Conservation and Society, 14*(4), 293–304.

Antoinette, J. (2012). *Speaking for the enslaved: Heritage interpretation at antebellum plantation sites.* Left Coast Press.

Borokini, T. (2016). Sanctuary of the spirits: *Okwu-muo, Ori Oke* and 'mammy water' in the veneration of sacred natural sites in southern Nigeria. *International Journal of Intangible Heritage, 11*, 56–70.

Brody, H. (1998). *Maps and dreams: Indians and the British Columbia frontier.* Waveland Press.

Carroll, C. (2014). Native enclosures: Tribal national parks and the progressive politics of environmental stewardship in Indian Country. *Geoforum, 53*, 31–40.

Cumberbatch, J. (2013). Barbadian bio-cultural heritage: An analysis of the flying fish. *International Journal of Intangible Heritage, 8*, 118–133.

Cummins, A., Farmer, K., & Russell, R. (Eds.). (2013). *Plantation to nation: Caribbean museums and national identity.* Common Ground.

Deloria, V. (1997). *Red earth, white lies: Native Americans and the myth of scientific fact.* Fulcrum Publishing.

Drengson, A., & Inoue, Y. (Eds.). (1995). *The deep ecology movement: An introductory anthology.* North Atlantic Books.

Geertz, C. (1973). *The interpretation of cultures.* Basic Books.

Gingrich, A. (2009). Evidence in socio-cultural anthropology: Limits and options for epistemological orientation. *The International Journal of Anthropology, 53*(2), 177–190.

Goldman, A. (1999). *Knowledge in a social world.* New York.

Gudykunst, W. (Ed.). (2003). *Cross-cultural and intercultural communication.* Sage.

Hall, E. (1959). *The silent language.* Anchor Books, Random House.

Holzman, D. (2012). Accounting for nature's benefits: The dollar value of ecosystem services environmental. *Health Perspectives, 120*(4), A53–A157.

Jackson, P. (1981). Phenomenology and social geography. *Area, 13*(4), 299–305.

Jett, S. (1992). The great race to "discover" Rainbow Bridge in 1909. *Kiva, 58*(1), 3–66.

Joosse, S., Powell, S., Bergeå, H., Böhm, S., Calderón, C., Caselunghe, E., Fischer, A., Grubbström, A., Hallgren, L., Holmgren, S., Löf, A., Nordström Källström,

H., Raitio, K., Senecah, S., Söderlund Kanarp, C., von Essen, E., Westberg, L., & Westin, M. (2020). Critical, engaged and change-oriented scholarship in environmental communication: Six methodological dilemmas to think with. *Environmental Communication, 14*(6), 758–771. https://doi.org/10.1080/17524032.2020.1725588

Keller, D. (2009). Deep ecology. In J. B. Callicott & R. Frodeman (Eds.), *Encyclopedia of environmental ethics and philosophy* (pp. 206–211). Macmillan Reference USA.

Lowie, R. (1937). *History of ethnological theory*. Facsimile Publisher. (in 2015).

Lynch, P. (1996). Students' alternative frameworks for the nature of matter: A cross-cultural study of linguistic and cultural interpretations. *International Journal of Science Education, 18*(6), 743–752. https://doi.org/10.1080/0950069960180607

Lyver, P. O.'. B. (2005). Co-managing environmental research: Lessons from two cross-cultural research partnerships in New Zealand. *Environmental Conservation, 32*(4), 365–370.

Maffie, J. (2000). Alternative epistemologies and the value of truth. *Social Epistemology, 14*, 247–257.

Maffie, J. (2020). Ethnoepistemology. *Internet Encyclopedia of Philosophy*. https://www.iep.utm.edu/ethno-ep/

Malinowski, B. (1922). *Argonauts of the western Pacific*. Routledge and Kegan Paul.

Malinowski, B., & Kayberry, P. (1946). *The dynamics of culture change*. Yale University Press.

Malinowski, B., Leech, E., & Berry, J. (1935). *Coral gardens and their magic*. Allen & Unwin.

Martinez, A. (2020). *Counterstory: The rhetoric and writing of critical race theory*. National Council of Teachers and English.

Murdock, G., Ford, C., Hudson, A., Kennedy, R., Simmons, L., & Whiting, W. (1982). *Outline of cultural materials* (new ed.). Yale University, Human Relations Area Files, Inc.

Ness, A. (1973). The shallow and the deep, long-range ecology movement, a summary. *Inquiry, 6*(1–4), 95–100. https://doi.org/10.1080/00201747308601682

Peterson, K. (2012). Ecosystem services, non-human agencies, and diffuse dependence. *Environmental Philosophy, 9*(2), 1–20.

Pitblado, B. (1998). Peak to peak in Paleoindian times: Occupation of Southwest Colorado. *Plains Anthropologist, 43*(166), 333–348.

Rappaport, R. (1999). *Ritual and religion in the making of humanity*. Cambridge University Press.

Rudolff, B., & al Zekri, M. (2014). A network of traditional knowledge: The intangible heritage of water distribution in Bahrain. *International Journal of Intangible Heritage, 9*, 84–97.

Ruppert, D. (2017). *Rainbow bridge traditional cultural property* (Listed as a TCP in the National Registrar of Historic Places). National Park Service.

Ruppert, D., & Smythe, C. (2017). National Park Service approaches to connecting indigenous cultural spiritual values to protected places. Chapter 7. In F. Sarmiento & S. Hitchner (Eds.), *Indigeneity and the sacred: Indigenous revival and the conservation of sacred natural sites in the Americas* (pp. 133–158). Berghahn.

Sessions, G. (1987). The deep ecology movement. *Environmental Review, 11*(2), 105–125.

Smith, L. (2012). *Decolonizing methodologies: Research and indigenous people.* Zed Books.

Soma, T. (2012). Contemporary falconry in Alta-Kazakh in western Mongolia. *International Journal of Intangible Heritage, 7*, 104–111.

Stoffle, R., Arnold, R., & Bulletts, A. (2016). Talking with nature: Southern Paiute epistemology and the double hermeneutic with a living planet. In G. Tully & M. Ridges (Eds.), *Collaborative heritage management* (pp. 75–100). Gorgias Press.

Stoffle, R., Halmo, D., Evans, M., & Olmsted, J. (1990). Calculating the cultural significance of American Indian plants: Paiute and Shoshone ethnobotany at Yucca Mountain, Nevada. *American Anthropologist, 92*(2), 416–432.

Stoffle, R., Sittler, C., Van Vlack, K., Pickering, E., & Lim, H. (2020a). Living universe or GeoFacts: Stone arches in Utah National Parks – Epistemological divides in heritage environmental communication. *International Journal of Intangible Heritage, 15*, 15–27.

Stoffle, R., Sittler, S., Pleshet, N., Alberte, M., Kays, C., Penry, G., Velasco, D., & Cachora, N. (2020b). *Ethnographic overview and assessment of Natural Bridges National Monument* (Interim Draft). University of Arizona.

Stoffle, R., Seowtewa, O., Kays, C., & Van Vlack, K. (2020c). Sustainable heritage tourism: Native American preservation recommendations at Arches, Canyonlands, and Hovenweep National Parks. *Sustainability, 12*(9846), 2–24.

Stoffle, R., Van Vlack, K., Arnold, R., & Bulletts-Benson, G. (2017). Cant of reconquest: Bringing American Indian voices into U.S. Southwest heritage management. In M. Albert, F. Bandarin, & A. Roders (Eds.), *Going beyond: Perceptions of sustainability in heritage studies* (pp. 231–246). Walter De Gruyter.

Stoffle, R., Zedeño, M., Eisenberg, A., Toupal, R., & Carroll, A. (2004). Shifting risks: Hoover Dam Bridge impacts on American Indian sacred landscapes. In A. Boholm & R. Lofsstedt (Eds.), *Facility siting: Risk, power, and identity in land use planning* (pp. 127–143). Earthscan.

Stoffle, R., & Zedeño, N. (2002). The concept of power in Numic and Yuman epistemology. *The High Plains Applied Anthropologist, 2*(22), 172–193.

Tilley, C. (1994). *Phenomenology of landscape places, paths, and monuments.* Berg.

Tilley, C. (2004). *The materiality of stone: Explorations in landscape phenomenology.* Berg.

Tilley, C. (2010). *Interpreting landscapes: Geologies, topographies, identities, explorations in landscape phenomenology*. Left Coast Press.

Toupal, R., & Stoffle, R. (2004). *Traditional resources of the Flagstaff Area Monuments*. Bureau of Applied Research in Anthropology, University of Arizona.

University of Oslo, Norway. (2020). Center for Development and the Environment. *UiO*. Retrieved December 2020 from https://www.sum.uio.no/english/

Varner, G. (2012). *Portals to other realms: Cup-marked stones and pre-historic rock carvings*. Lulu Press.

Winstanley-Chesters, R., & Ten, V. (2016). New goddess at Paektu Mountain: Two contemporary Korean myths. *South/North Korean Humanities, 2*(1), 151–179.

Yearly, S. (2006). How many "ends" of nature: Making sociological and phenomenological sense of nature. *Nature and Culture, 1*(1), 10–21.

“The Sea Has No Boundaries”: Collaboration and Communication Between Actors in Coastal Planning on the Swedish West Coast

Simon Larsson and Annelie Sjölander-Lindqvist

Introduction

Confronted with the challenges of coordinating spatial planning of the Swedish West Coast, regional actors have initiated a project to increase the degree of collaboration between societal actors in coastal planning.[1] The project was initiated by regional organizations, but planners, business

[1]The County Administrative Board (CAB) of Västra Götaland, the Västra Götaland Regional Council (VGR), the Gothenburg Region (GR), and Business Region Gothenburg (BRG).

S. Larsson (✉)
Gothenburg Research Institute, The Gothenburg School of Business, Economics and Law, University of Gothenburg, Göteborg, Sweden
e-mail: simon.larsson@gri.gu.se

A. Sjölander-Lindqvist et al. (eds.), *Anthropological Perspectives on Environmental Communication*, Palgrave Studies in Anthropology of Sustainability, https://doi.org/10.1007/978-3-030-78040-1_8

developers, and environmental managers working in eight coastal municipalities participated, as well as representatives from interest organizations. The project, which lasted from 2016 to 2019, was aimed at dealing with the main problems and constraints facing the sustainable development of coastal and marine areas, and strengthening cooperation and establishing common objectives regarding the planning of a coastal zone that includes the geographical area of the involved municipalities.[2]

In the context of this chapter, this collaborative project is studied as an example of environmental communication within and between organizations. The broad and complex set of socio-environmental challenges facing society requires awareness of the processes of environmental communication; here, the anthropological tradition of interpretation and of stressing how communication is "embedded in social institutions, local practices and experiences and is understood and judged in terms of emic, localized, collectively defined values and concerns" (Boholm, 2015, p. 158) can make a significant contribution.

This chapter focuses on environmental communication as a continuous interaction between actors in a specific institutional setting. Such collaborations and communication are not always straightforward, nor consensus-driven or devoid of conflicts, since the different actors have their own interests and perspectives. In this chapter, we consider the complexities associated with inter-agency dialogue processes and the difficulties involved in developing and coming to agreements about plans conceived to cope with environmental issues.

Our study draws on the idea that collaborative efforts and dialogue are concerted by fortuitous and lived practices and affected by contextual value-driven circumstances, such as sector-specific tools, regulations and policies, resources, and norms (c.f., Hammond & Brandt, 2004; Shore

[2] The project covers the geographical area of the following municipalities on the Swedish West Coast: Uddevalla, Orust, Tjörn, Stenungsund, Kungälv, Öckerö, Göteborg, and Kungsbacka.

A. Sjölander-Lindqvist
School of Global Studies & Gothenburg Research Institute,
University of Gothenburg, Göteborg, Sweden
e-mail: annelie.sjolander-lindqvist@gu.se

et al., 2011), which, we argue, need to be acknowledged to reach a more profound and critical understanding of communication as an institutionalized practice. Therefore, we focus on the collaborative process as a shared space for human experience, imagination, and significance. With this as our departure point, we propose that any critical analysis of collaborative environmental communication must take the framework of institutional(ized) roles and responsibilities in consideration as it creates the framework in which communication takes place (Sjölander-Lindqvist, 2015; Sjölander-Lindqvist et al., 2020).

We explore the communication that is taking place as an ongoing negotiation of interests and responsibilities. We utilize the concept of "boundary work" (Gieryn, 1983, 1999) in our exploration of the multidimensional character of communicative practices and contents as they unfolded in the project. We followed the project *Inter-Municipal Coastal Zone Planning: In the Gothenburg Region, Orust och Uddevalla* for a year and a half to explore how the involved actors understood the objectives of the collaboration and to identify the driving forces and obstacles for fruitful cooperation, communication, and goal completion. Through interviews and by participating in meetings and discussion sessions, we gained insight into how the project was situated within the institutional logics of the involved actors' organizations, explored how they understood the objectives of the collaboration, and identified the driving forces and obstacles for fruitful cooperation and communication between the concerned actors.

Through data collection, which included approximately 50 hours of participant observations in meetings, 13 formal recorded in-depth interviews with people associated with the project, informal conservation, as well as documentation from the project available on the webpage of the Gothenburg Region (GR), we captured how the project participants made sense of the project and the meanings attached to the particularities and circumstances of the collaboration. Environmental communication in this regard is studied from an anthropological perspective, highlighting the added value of doing participant observations and focusing on how emic models of explanations are (hierarchically) structured according to institutional roles and responsibilities of the addressees of the communication. This study demonstrates that some aspects of a collaborative project only become possible to observe if one studies it for a longer duration, becoming a natural element in the environment and getting to know the participants.

Theoretical Context

Environmental communication does not only take place between governmental bodies, business companies, and the public. There is also a continuous dialogue within and between government authorities, regional and municipal bodies, and other stakeholder organizations (e.g., Larsson et al., 2019; Sjölander-Lindqvist, 2015). This communication is situated within the institutional logics of involved organizations, and actors will follow their institutional logic and make decisions accordingly (Luhmann, 1989). However, institutionalized attitudes or patterns of behavior do not always follow a rationale that makes sense in relation to specifically defined goals within the organization. Attitudes and patterns of behavior might also be dependent on institutional praxis and be based on contingent factors such as influential individuals within the organization and an organizational culture that has developed and become institutionalized over time. A focus on the processes and the ongoing negotiations of meaning is essential for understanding the complex "nature" of communication within an institutional setting.

Differences in perspectives and conflicts between societal actors can relate to certain specific issues, but it can also connect to knowledge claims and disputes over areas of responsibility between societal bodies (Boholm & Larsson, 2019; Sjölander-Lindqvist et al., 2020). These conflicts and negotiations can be described as *boundary work* in Gieryn's (1983) wording. *Boundary work* is a concept initially used to describe how divergent knowledge claims are made and negotiated between different scientific fields (Gieryn, 1983, 1999), but the concept has also been utilized to analyze how organizations negotiate the scope of their organizational responsibilities and knowledge claims (e.g., Zietsma & Lawrence, 2010). From such a perspective, organizations or subdivisions of organizations might compete over claims to responsibilities and jurisdictions with a particular issue or domain as well as the authority to formulate the problem, but they might also try to push away such responsibilities.

Diverging organizational logics and perspectives of the involved actors (i.e., values, interpretative schemata, and assumptions) become particularly evident during a reorganization, or when actors are to collaborate over established institutional boundaries. The inter-municipal project discussed in this chapter is, therefore, a telling example that we can use to discuss the complexities associated with inter-agency collaboration and communication.

The Rationale for the Collaborative Project

A frequently touted phrase in interviews with participants in the project and the documentation of the collaborative project on coastal planning and management is that *the sea has no boundaries*. This truism is understood as an imperative for collaboration over administrative and geographical boundaries as regarding the planning and management of the sea and the coastal zone.[3] Coming as a researcher from the outside, it is not easy to understand why the sea, in contrast to land, is considered to be without borders or boundaries, or for that matter why this is a call for cross-sectoral collaboration—but let us leave this question aside and take a step back.

In Sweden, municipalities have a far-reaching responsibility for planning and granting building permits within their geographical borders (referred to as "the plan monopoly"). The Swedish Code of Statuses 2010:900 stipulates that each municipality must have an updated comprehensive spatial plan covering the entire geographical area of the municipality. It should reflect the opinion of the political majority and be ratified by the municipal council. The spatial plan is not a binding document, but should guide decisions on how land and water areas should be used and managed, and how the built environment can be used, developed, or preserved. For the coastal municipalities, this also includes the responsibility of establishing maritime spatial plans up to 12 nautical miles from the baseline[4] (these plans are also sometimes referred to as "blue comprehensive plans"[5]).

More often than not, coastal municipalities lack comprehensive maritime spatial plans; this can be explained by both a lack of political interest (reflecting a lack of interest among the inhabitants) and a lack of resources for doing this work, especially within the smaller municipalities. A respondent from one of the regional organizations said in an interview, "The sea is a low priority area, reflecting a limited interest in these issues among the

[3] For example, "The boundless character of the Sea calls for a holistic view stressing collaboration over different sectors as well as administrative and geographical boundaries" (Förstudie mellankommunal kustplanering i Göteborgsregionen, Orust och Uddevalla https://goteborgsregionen.se/download/18.4e85091914e1a80913d5e985/1435055444244/Remissversion+-+Mellankommunal+kustplanering+i+G%C3%B6teborgsregionen%2C+Orust+och+Uddevalla.pdf). All translations from Swedish to English are by the first author of this chapter unless otherwise noted.

[4] The baseline is the line outside the islands.

[5] Note that this is not a formal legal term.

majority in the municipalities. That makes it difficult to spend resources on this." Although the municipalities have had the authority and responsibility to develop the comprehensive blue plans, very few municipalities have put any effort into this since the Swedish Code of Statuses came into force in 1987.[6] In 2010, only 4 out of 84 Swedish coastal municipalities had such plans. Referring to the lack of maritime regulation and planning, one informant said that "the sea is the last wild west."

In the early 2010s, a dialogue regarding the planning of the coastal area between regional actors was initiated, including the County Administrative Board (CAB) of Västra Götaland, the Västra Götaland Regional Council (VGR), the Gothenburg Region (GR), and Business Region Gothenburg (BRG). A jointly conducted pre-study, carried out in 2014–2015, identified the need for collaborative planning efforts and a need to coordinate the national, regional, and municipal responsibilities for the sea.[7] The preliminary study proposed a collaborative project involving the six coastal municipalities within the Gothenburg Region, complemented by the two municipalities Uddevalla and Orust. One model for this collaborative project was a similar collaborative project conducted at an earlier time by the coastal municipalities in northern Bohuslän, a nearby West Coast province. The preliminary study formulated a "proposal on how future in-depth cooperation on inter-municipal coastal planning can be developed, concerning priority thematic areas." This study created the guidelines for the project discussed in this chapter.

The project did not materialize in a political vacuum; the involved actors relate this project to ongoing national planning for the future exploitation of Swedish marine resources, and as an EU member state, Sweden is obliged under the MSP Directive to develop a national maritime spatial plan by March 31, 2021, at the latest.[8] The collaboration project is also related to a trend in political governance stressing collaborative efforts in the resource management as a desirable, or even essential, way of ensuring sustainability when addressing environmental challenges (Hayes & Persha, 2010). It is, in general, assumed that increased participation and collaboration will lead to increased legitimacy, effectiveness, and sustainability (Emerson et al., 2012). Hence, the deliberative approach

[6] *Plan och bygglagen* first came into being in 1987 and was replaced by the current Swedish Code of Statuses in 2011.

[7] There is a geographical overlap between the municipal and national plans for the sea.

[8] https://www.msp-platform.eu/msp-eu/introduction-msp

to governance is in contemporary political and academic discourse touted to provide more socially efficient and robust decisions (e.g., Barber, 1984; Emerson et al., 2012; Hansson-Forman et al., 2018). Based on the normative ideals of shared responsibility and mutual learning, this form of governance proves useful in handling conflicting goals and seeking acceptable and legitimate outcomes of decision making (Agrawal & Ribot, 1999).

Furthermore, interaction across sectors or pooling of knowledge spheres can promote the development of new knowledge (Mårald et al., 2015) which may also be more sensitive to the given context and to the specificities of the place (Sjölander-Lindqvist & Cinque, 2014). Our case for interrogation reveals how the project is rationalized against the contemporary currency of a collaborative norm for policy work by the inclusion of not only municipal actors but also several other societal stakeholders.

The Organization and the Explicit Objectives of the Project

It was decided at an early stage that the GR[9] was going to host the project. Still, the project was funded by the VGR, the Swedish Agency for Marine and Water Management (funds distributed through the CAB), BRG (through in-kind), as well as by the participating municipalities.[10] The project lasted between 2016 and 2019. The participants from the municipalities were civil servants who worked with environmental issues, planning, and business development. In addition to the organizations mentioned earlier, various interest organizations were also actively involved in the project.[11]

The explicit aim of the project was to establish intensified collaboration between municipalities, authorities, and other relevant actors regarding social, cultural, and environmental planning. In other words, it aimed at including both horizontal collaboration (between municipalities) and vertical collaboration (between the municipalities and the regional actors) across administrative and geographical boundaries. The primary outcome of the project was to deliver a "structural picture" (*strukturbild*) with joint

[9] The GR is a cooperative organization uniting some 13 municipalities in western Sweden (6 of these municipalities border the sea).

[10] The participating municipalities are listed in Footnote 1.

[11] Västarvet, Västkuststiftelsen, Maritima klustret, 8 Fjordar, Bohusläns skärgårdsråd, Turistrådet Västsverige, Vattenrådet för Bohuskusten, Göta Älvs Vattenråd.

agreements on how the coastal zone should be developed and preserved sustainably. This agreement was to be ratified in the GR Council and the municipality executive boards or municipality councils in each participating municipality.[12] This meant that the document, although not legally binding, was expected to have some political weight. Another expected outcome was a knowledge platform to assist the municipalities' spatial planning. This was to be presented in a joint digital GIS[13] platform that could provide relevant planning information such as the location of green areas, nature reserves, and other infrastructure.[14] This information, to be included in the GIS database, was to be collected from the individual municipalities, government authorities, and interest organizations. A consultancy agency was also hired to produce three reports on (i) accessibility to the coast, (ii) marinas, and (iii) production of energy at sea. Information from these reports was to feed into both the structure plan and the GIS database.

Project leaders from the GR ran the project in dialogue with a strategic leadership group comprising representatives from the CAB of Västra Götaland, VGR, BRG, and GR. Throughout the project, there were joint meetings with all participants, but most of the work was carried out in four working groups: *Structure*, *Utilization of the Sea*, *Experiences*, and *Environment*.[15] These groups had monthly meetings for the initial two years of the project and included civil servants representing the involved municipalities (working in planning, environment, and business development), representatives from the GR, and participants from other interest and agency organizations operating in the coastal zone. Occasionally, the CAB and VGR participated in the meetings to discuss particular issues.

The working groups produced preliminary reports on their topic. Although the topics for the working groups had been defined, they had a large degree of autonomy and flexibility in dealing with the topics and deciding what to include in their reports. For example, the *Experience* working group added tourism and recreation to their list of topics although

[12] This varies among the involved municipalities.

[13] Geographic Information System.

[14] Now available online at http://karta.goteborgsregionen.se/

[15] Initially the project identified nine themes to work with: (1) aquaculture and fisheries, (2) access to the coast, (3) boating, (4) transport infrastructure, (5) visitor industry, (6) energy production at sea, (7) environmental state of the sea, (8) drilling sites, and (9) principles of construction. For practical purposes, these themes were merged into four working groups.

it was not included in the initial plan. The working groups compiled existing documentation and identified needs or knowledge gaps within their area of responsibility—information that could feed into the GIS platform. The working groups also invited external lecturers from the different interest organizations, for example, to support the building of knowledge and increase the participants' competence but also to obtain information that could feed into the reports to be produced by the groups. Each group presented the results in the form of a preliminary report that would feed into the final structural picture.

When the working groups were ready with their preliminary reports, an operational group was formed. It was led by the project manager from the GR and included representatives from each working group. The operational group produced the structural picture from the reports in each working group and the material provided by the consultancy agency; they did this in dialogue with the strategic management group as well as with the politicians in the municipalities.

The Benefits of the Project from the Perspective of the Involved Stakeholders

From the perspective of the participating municipalities, the project was relevant to the everyday work of the involved departments as well as to their planning responsibility. Matters related to the sea are especially relevant to the Environment Department, the Planning Department, and the Department for Business Development. However, there are significant differences between what topics are of relevance to the involved municipalities. For example, the municipality of Kungälv has problems with the protection of eelgrass meadows, for Uddevalla the issue of the harbor is important, and for Orust aquaculture and fishing is of the essence. The VGR and BRG are mainly involved in issues concerning business development when it comes to issues related to coastal planning. The CAB has an interest in development as well as environmental protection, and they stressed that they want to contribute to the project's implementation with their knowledge and established contact network. The other interest organizations were involved insofar as it related to their specific roles, including business development, environmental protection, and cultural heritage preservation.

Although the concerned actors had different perspectives and different roles, there was a general understanding among the participants that some matters needed to be coordinated between the municipalities as well as between the regional actors. A civil servant working with business development in one of the municipalities said in an interview:

> For example, we must coordinate marine aquacultures, and shipping must be discussed with our neighboring municipalities [...] It doesn't work if [our municipality] is just looking at our own problems [...], so I think it's obvious that we need to collaborate and coordinate the work between the municipalities.

Areas considered essential to collaborate on included public transportation, environmental pollution, construction of marinas, and shipping. In fact, all participants, to some degree, stressed the benefits of joint planning and collaboration.

The civil servants working in the municipalities' planning departments particularly highlighted the benefits of the project. One planner said that all the issues dealt with in the project would be used in their plans, stating:

> [T]he structural picture points out sites for aquaculture, shellfish farms, and everything else, how people get to the coast, and how they use the coastal area. It'll be a basis for us to do our planning. We'll be able to see what the interests are; it's like any other planning document, really.

Primarily, the GIS platform was understood to be a useful tool in making the overview plans for the coastal areas in the municipalities. Another civil servant in a municipality stated:

> The planning documentation that we have received [through the collaborative project] gives us a whole GIS product, where we can just put all the available information and copy–paste; a lot of the planning has already been done by other actors, so we can do our planning in relation to the documentation that already exists.

The civil servants from the involved municipalities saw benefits in aligning their municipal planning with the project's regional plans. However, the planning process in the municipality was not only understood to benefit from the outcome of the project, but also from the conversations and discussions in the working groups and joint meetings. One of the municipal planners said:

> And of course, we talk about important issues in the different working groups, concerning how we should think about energy supply and all these issues. It will lend an increased depth to the planning documentation compared to if I had to do everything by myself.

The informants understood the group discussions and the joint meetings within the project as beneficial to their work. They deepened their understanding of the issues and the values at stake, and their ability to "see how the different interests are interconnected." Furthermore, the collaborative project is understood as helpful in making the overview plans in the municipalities.

In addition to the benefits of the expected outcomes of the project, it is also stressed that they get to know civil servants in the other municipalities, and people within the interest organizations. A civil servant from a municipality said that it "has been incredibly valuable to get all the contacts I have got, we have met a lot of people, and widened the contact areas between municipalities."

The element of learning was also emphasized in several interviews. For example, a civil servant said:

> We have different traditions and different experiences, and I think the coastal zone project has been great in that you've had to listen to different municipalities [...] if you take Orust and Tjörn as examples, they have a completely different tradition with the fishing industry, compared to us in Uddevalla. But it's important that we're filled in on what the other municipalities are doing, so that we don't contribute to anyone else's problems.

One civil servant told us that he was recently hired by the municipality and that he had worked his whole life in the business sector:

> So, for me [the collaborative project] has been a part of my process of understanding what my role is as a business developer [in the municipality]. I have no experience whatsoever with comprehensive planning; all these processes were completely new to me.

In other words, he understood the project as a way of understanding the public sector better. Another civil servant from a municipality stressed that this learning process was more important than the actual outcome of the project:

> For my part, I've learned a whole lot of things that I wasn't aware of before thanks to these networks with other people who are professionals in their fields. Eelgrass meadows are one example, marinas, ferries, aquaculture, and I really got to learn a great deal, things that might also be beneficial to [...] municipal planning in the future.

As we have seen, the benefits of the project were formulated as aligned with the project's overtly formulated goal, but there were also benefits in terms of learning and networking.

Conflicts of Interest

Just as there was a shared understanding of the benefits of the project, the informants also brought up difficulties in making joint agreements. There were strong opinions and diverging interests regarding how the coast and the sea should be utilized, and there were also some conflicting interests and values at stake. For example, renewable energy solutions such as offshore wind farms might infringe on the space of other interests such as the military or ship traffic. Hence, the development of one sector may create problems in another sector. Most conflicts discussed in the collaborative project related to environmental protection versus some other perceived benefit such as exploitation of new land or business development. An example of one such conflict is when the representative from BRG in a joint meeting raised his concern that increased protection for the eelgrass meadows in the structural picture would heavily influence business development, saying that current regulation already provides "extremely" strong protection of the eelgrass. An agitated discussion followed this statement, in which the civil servants working with environmental issues argued that the eelgrass did not enjoy "extreme" protection, and if that had been the case it would not have decreased by 60% in recent years.

Conflicts of interest were most clearly articulated in the discussion on commercial fishing. Whereas the civil servants from the municipalities' environmental departments stressed how the fish stock and marine environment are negatively impacted by the fishing industry, the business side stressed that the fisheries constitute the fishermen's livelihood, and that this is important also for retailers and the local fish-processing industry. A vibrant fishing industry, they continued, is also an important part of attracting tourism to the area. The fisheries are also understood to be an essential aspect of the identity and cultural heritage of the coastal area in

Sweden, embedded in cultural values and the identities of many fishing societies. The disagreements mostly related to the difficulties in balancing economic, social, and ecological interests, but sometimes the conflicts were also about how to interpret information and evaluate information from different sources.

What, then, were the dividing lines between the various positions within the project? A project manager said that the initial problem was that everyone was used to thinking within the established boundaries of their organizations and that the municipalities were used to dealing with issues at the municipal level and promoting their own local interests. Nevertheless, the discussions showed that there were different positions within the municipalities. For example, two civil servants from the same municipality had entirely different views on fishing. One believed that all fishermen in Sweden should be paid salaries for the rest of their lives to discontinue fishing, while the other understood fishing as an essential part of the municipality's survival and identity. It may be needless to say that the civil servants that wanted to discontinue fishing worked in the environment department, while the person promoting the fishing industry worked in the department for business development.

Obstacles Related to Regulations and the Institutional Division of Responsibilities in Society

While nearly all the involved actors stressed the benefits of joint agreements, it was very difficult to come to agreements when it came down to specific issues. While these difficulties sometimes related to conflicts of interests between the involved actors, many obstacles in formulating joint goals related to existing principles for making decisions and the formal roles and responsibilities of the concerned actors. One prominent example of this is when one of the working groups wanted to coordinate the disposal of mud from dredging and propose suitable locations for such disposal sites. They discussed this issue and the criteria for suitable sites for a long time before the CAB communicated that this was not possible under current legislation.[16] Disposal of mud is not permitted by Swedish regula-

[16]According to the Environmental Code (Milijöbalken), Chapter 31, Section 31, no one may dispose of mud or any other waste within Sweden's maritime territory and economic zone.

tions. While exceptions can be made, the CAB must receive a formal application each time and cannot designate sites for disposal.

The working group did not accept this at first and decided to create their own interpretation of the existing regulation in order to find loopholes that would allow the municipality to point out what they considered as proper sites for mud disposal. After this discussion had proceeded for some time, one of the participants pointed out that none of them were legal experts and that they might not be able to draw any conclusion from their readings. Eventually, they came to the conclusion that within the project they did not have any ways to influence these decisions, which are made in each specific case, and in accordance with specific regulations and guidelines. An informant within the project said that they "[t]hought it would be possible to point out suitable places from a joint regional perspective, but you really can't do that in this project." This is one example of how actors in the project tried to expand the scope of their role and responsibility, while the CAB, on the other hand, policed the borders of their responsibilities and jurisdiction.

There were several obstacles within the project that related to the confusion of roles and responsibilities in society. Another example was when the working groups were to formulate the wordings of their reports. In many instances, the working groups wanted to make statements in the report that advocated further-reaching environmental protection than that stipulated by contemporary regulation. This was often influenced by the participants' ideological standpoints. On this subject, one participant said during a meeting that "we have to look upon this as a long-term challenge. No one is speaking in the interest of nature, [...] we have to stand up and speak for interests that don't have their own voice." The shoreland was one area where the working group wanted to make protection stricter than existing regulations. One group also proposed that farmland should not be further exploited, motivated by the opinion that it is "unreasonable to transport food across the globe" for environmental reasons—quite contrary to contemporary politics and policies.

While a few of these formulations made it all the way into the report (e.g., the issue concerning farmland), some of the proposed wording in the preliminary reports were too strong for the politicians in the municipalities to stomach. This was especially noticeable with regard to the suggested increased protection of shoreland, which the politicians from the municipalities did not want to ratify. A liberal view of utilizing the shoreland is seen as an important aspect of making the municipality more attractive to investment (e.g., making it possible for people to build summer houses close to the sea). In the final version of the structural picture, large changes had been made to what had been decided in the working group. The project management concluded that "[t]he visions had to be formulated very broadly for all the involved actors and municipalities to accept it."

In all these examples, the working groups tried to transcend the institutional boundaries and jurisdictions of the involved organizations, something that can be described as boundary work (as described earlier), where actors try to position themselves and extend the reach of their jurisdiction (or patrol and protect its borders). Some of these examples can be understood as resulting from a lack of knowledge about regulations and institutional responsibilities among the participants, but they can also be interpreted as active expressions of wanting to voice an opinion or trying to extend the reach of one's own jurisdiction and, in so doing, trying to expand the boundaries of one's area of responsibility.

The concept of intuitional boundaries is essential to understanding the layers of discourse in the communication within the project. When a person makes a statement in the working group, their affiliation and position within that organization is of the essence for interpreting and juxtaposing the layers of discourse. When discussing the communication in the collaborative project as boundary work between different actors, it is important to acknowledge not only existing, established actors but others as well; a temporary unit within the collaborative project like a working group (or, for that matter, the collaborative project in its entirety) can create an institution of its own, operating with its own institutional logics, especially if the project exists for a longer period of time.

While an initial problem was described by the project management as being related to how the involved actors were used to thinking within the traditional boundaries of their own municipalities and organizations, a difficulty arising later on in the project was that it was too difficult to "think inside" the framework of existing institutional boundaries (as demonstrated by the examples above). It was also confirmed in interviews that the project participants' attitudes differed from those of civil servants that did not engage in the project—so that participants in the project established joint viewpoints and joint objectives in contrast to those who did not participate in the project within the municipalities and the other organizations. Consequently, the positions within the project were not something given, but rather under continuous negotiation. Actors might side with other actors in discussions dependent on a number of factors, and their position was not given beforehand, but depended on contingent factors such as group dynamics. So while the project aimed at transcending certain boundaries between organizations, it simultaneously created new boundaries within the organizations; this is an insight that is important to take into consideration when designing collaborative projects.

Communication Through Friction

Taken together, the working group spent weeks and months trying to reach agreements about things that were not possible due to current laws and a regulated division of responsibilities between societal institutions. Much work was put into the project groups, and the results were meager in terms of their outcome in the final joint agreements. To outside observers, some of these discussions seemed fruitless, and we expected the participants to be more negative toward the collaborative project and to view it as a waste of time and something that would potentially aggravate the participants and increase polarization. And furthermore, we expected that the actors would be disappointed or disillusioned when the agreed-upon solutions to the issues discussed were not possible to realize.

There were also some complaints from the participants who shared this line of reasoning. One of the planners said during a coffee break that the work to establish joint agreements was unnecessary because these visions were too difficult to integrate with her everyday work, where she had to rely on current regulation and planning documents. A civil servant argued that the collaborative project would have benefited from firmer leadership and said that a clearer framework for the collaboration and discussions should have been formulated from the very beginning of the project. Another civil servant said in an interview, "I think the work of the working groups could have been organized a little differently; I think we should have worked more with the vision and purpose of the project at the beginning to determine what we wanted to achieve within the project." Another person stressed that it should have been clearer from the beginning what was possible to do within the framework of the project.

However, the participants were remarkably patient, tolerant, and in good spirits. Also, misunderstandings and conflicts were understood as a part of the learning process. In the interviews with the CAB, they recognized the difficulties facing the project and the lack of knowledge among the participants in the municipalities, which led to conflicts over roles and responsibilities. They did not, however, view this as an obstacle for collaboration. Rather, they claimed that these conflicts were a part of a learning process. They said that it was beneficial to let the municipalities discuss quite freely and intervene only when they saw it as necessary. Intervening in the collaborative project was something the CAB understood as allowing them to educate the municipalities and stop ideas at an early stage, making them less prone to making uninformed decisions in the future. "You can punch holes in their bubbles before they become big balloons ... you can deflate [them] slowly instead of punching a hole." So while the municipalities saw the project as a collaboration between actors, the CAB also understood the objective of the project to be educating the municipalities and encouraging them to take on their responsibilities in issues related to planning.

From this discussion, we can see that the actors involved in a collaborative project can have intentions that diverge from its formulated goals, and that a collaborative project may have perceived benefits that are not immediately evident in its overtly defined objectives. In relation to environmental communication, it is interesting to see that what appeared as misunderstandings and conflicts was understood by some actors as a good way to communicate about environmental laws and regulations.

Concluding Discussion

The stakeholders involved in this project agreed that several issues were conducive to being addressed on a regional level in collaboration with other regional stakeholders. However, they also recognized that collaboration is difficult to achieve in practice. Through several examples, this chapter has shown that the established institutional division of roles and responsibilities, as well as current legislation, creates challenges for cooperation and producing outcomes in a collaborative project. While the sea is described by the actors in the project as having no borders or boundaries, there clearly are boundaries in the political institutions and legal framework guiding the collaborative process—which limit the possibility of making decisions in a collaborative project involving actors from different organizations. The involved actors had diverse rationalities aligned with their roles and responsibilities, and their actions and standpoints were guided by diverse ways of assessing the problems and specific legislations. At the same time, the collaborative project established new viewpoints and new ways of doing things. Boundary work is a fruitful perspective for conceptualizing the dynamics and ongoing communication within a collaborative project, viewing the project as a way of negotiating roles and responsibilities or maintaining a certain order. This chapter has also stressed that the collaborative project was understood to have several positive outcomes not formulated directly in its deliverables, and also that the friction, misunderstandings, and conflicts created when trying to transcend a project's boundaries can be understood as a positive outcome of a project by providing opportunities for involved actors to learn.

In previous research on collaboration, it has been argued that there is a lack of empirical studies that follow collaborative projects for a longer time (San Martín-Rodríguez et al., 2005). Many studies of collaborative projects are made after the collaborative project is finished. Being an empirical study of a collaborative process, this chapter contributes to this scholarly field by stressing the benefits of utilizing social anthropological methods and perspectives in analyzing environmental communication in collaborative projects.

In this chapter, we have tried to show that a collaborative project has more dimensions than what can be evaluated by measuring the degree to which the goals of the projects are fulfilled or through some external measurements. A social anthropological perspective on policy work is well suited to investigating communication within a project, as it seeks to describe and understand the concerned parties' conditions for existence and working by exploring the meanings congregated around policy implementation, organizational relationships, and conceptual structures (Shore et al., 2011). In this study, we have utilized anthropological methods of participant observations in studying collaborative efforts and communication—and demonstrated how an anthropological focus on the working beliefs of any human organization can be fruitful in analyzing environmental communication.

Acknowledgments The authors would like to express their gratitude to Kathleen Van Vlack, Michael E. Dove, as well as the anonymous reviewers for providing valuable feedback on the previous drafts of this chapter. Thank you also to all the civil servants within the collaborative project who contributed their time and knowledge. Finally, we would like to express our gratitude to the Centre for Sea and Society at the University of Gothenburg for financing the study.

References

Agrawal, A., & Ribot, J. C. (1999). Accountability in decentralization: A framework with South Asian and West African environmental cases. *The Journal of Developing Areas, 33*(4), 473–502.

Barber, B. (1984). *Strong democracy: Participatory politics for a new age*. University of California Press.

Boholm, Å. (2015). *Anthropology and risk*. Routledge.

Boholm, Å., & Larsson, S. (2019). What is the problem? A literature review on challenges facing the communication of nanotechnology to the public. *Journal of Nanoparticle Research, 21*(86). https://doi.org/10.1007/s11051-019-4524-3

Emerson, K., Nabatchi, T., & Balogh, S. (2012). An integrative framework for collaborative governance. *Journal of Public Administration: Research and Theory, 22*(1), 1–29.

Gieryn, T. F. (1983). Boundary-work and the demarcation of science from non-science: Strains and interests in professional ideologies of scientists. *American Sociological Review*, 781–795.

Gieryn, T. F. (1999). *Cultural boundaries of science: Credibility on the line*. University of Chicago Press.

Hammond, L., & Brandt, C. (2004). Science and cultural process: Defining an anthropological approach to science education. *Studies in Science Education, 40*(1), 1–47. https://doi.org/10.1080/03057260408560202

Hansson-Forman, K., Reimerson, E., Sjölander-Lindqvist, A., & Sandström, C. (2018). Governing large carnivores—Comparative insights from three different countries. *Society & Natural Resources, 31*(7), 837–852.

Hayes, T., & Persha, L. (2010). Nesting local forestry initiatives: Revisiting community forest management in a REDD+ world. *Forest Policy and Economics, 12*(8), 545–553.

Larsson, S., Jansson, M., & Boholm, Å. (2019). Expert stakeholders' perception of nanotechnology: Risk, benefit, knowledge, and regulation. *Journal of Nano Research, 21*(3). https://doi.org/10.1007/s11051-019-4498-1

Luhmann, N. (1989). *Ecological communication*. University of Chicago Press.

Mårald, E., Sandström, C., Rist, L., Rosvall, O., Samuelsson, L., & Idenfors, A. (2015). Exploring the use of a dialogue process to tackle a complex and controversial issue in forest management. *Scandinavian Journal of Forest Research, 30*(8), 749–756.

San Martín-Rodríguez, L., Beaulieu, M. D., D'Amour, D., & Ferrada-Videla, M. (2005). The determinants of successful collaboration: A review of theoretical and empirical studies. *Journal of Interprofessional Care, 19*(supl), 132–147.

Shore, C., Wright, S., & Però, D. (Eds.). (2011). *Policy worlds: Anthropology and analysis of contemporary power*. Berghahn.

Sjölander-Lindqvist, A. (2015). Balancing differentiated interests and conceptualizations in environmental management: Working across sectors in Swedish river restoration. *Journal of Organizational Ethnography, 4*(3), 306–323. https://doi.org/10.1108/JOE-07-2014-0022

Sjölander-Lindqvist, A., & Cinque, S. (2014). Dynamics of participation: Access, standing and influence in contested natural resource management. *Partecipazione e Conflitto, 7*(2), 360–383.

Sjölander-Lindqvist, A., Larsson, S., Fava, N., Gillberg, N., Marcianò, C., & Cinque, S. (2020). Communicating about COVID-19 in four European countries: Similarities and differences in national discourses in Germany, Italy, Spain, and Sweden. *Frontiers in Communication: Political Communication and Society.* (Accepted for publication, pre-print available online). https://doi.org/10.3389/fcomm.2020.593325

Zietsma, C., & Lawrence, T. B. (2010). Institutional work in the transformation of an organizational field: The interplay of boundary work and practice work. *Administrative Science Quarterly, 55*(2), 189–221. https://doi.org/10.2189/asqu.2010.55.2.189

Power, Conflicts, and Environmental Communication in the Struggles for Water Justice in Rural Chile: Insights from the Epistemologies of the South and the Anthropology of Power

Cristian Alarcón Ferrari

Introduction

This chapter aims at advancing a conflict- and power-oriented conceptualization of environmental communication to analyze and explain struggles for water justice in Chile. In doing so, the chapter draws insights from the epistemologies of the South and the anthropology of power to more deeply understand environmental communication processes and to highlight the explanatory and analytical potential of a critical conceptualization

C. Alarcón Ferrari (✉)
Department of Urban and Rural Development, Swedish University of Agricultural Sciences, Uppsala, Sweden
e-mail: cristian-alarcon.ferrari@slu.se

A. Sjölander-Lindqvist et al. (eds.), *Anthropological Perspectives on Environmental Communication*, Palgrave Studies in Anthropology of Sustainability, https://doi.org/10.1007/978-3-030-78040-1_9

of environmental communication (see also Alarcón, 2015, 2019, 2020). The chapter builds and elaborates upon insights into power and communication developed by Paulo Freire and Eric Wolf to dialectically approach issues of hegemony, ideology, and discourses, and to more deeply address power relations and conflicts in the conceptualization of environmental communication. I argue that making sense of those dialectical interrelations serves to advance a critical conceptualization of environmental communication in both theoretical and empirical terms. Closely linked to a critical conceptualization of environmental communication, the chapter elaborates some normative perspectives for a situated understanding of environmental communication which builds upon Freire's engaged theoretical approach to communication, developed in the context of past political struggles in Chile. Empirically, the chapter focuses on the analysis and explanation of struggles for water justice and water democracy in Chile, providing empirical insights from fieldwork conducted in three administrative regions of Chile, including interviews, observations, and analysis of documents in Southern and Central Chile in recent years.

Environmental Communication from a Conflict- and Power-Oriented Perspective

To start with, I would like to stress that the relevance of a conflict- and power-oriented conceptualization of environmental communication arises from the fundamental place that this form of communication occupies in the current struggles around the present and the future of human interactions in ecosystems. Within this context, a sound theoretical approach to environmental communication has the potential to contribute to addressing in analytical and explanatory terms what Joas and Knöbl define as the three specific questions in the social sciences, namely: What is action? What is social order? And what determines social change? (Joas & Knöbl, 2009). Furthermore, I maintain that a critical conceptualization of environmental communication can bring new theoretical and normative perspectives to critical theories trying to better understand the fundamental role of power in today's specific social-ecological conflictivity. Here it is important to consider that there has long been an ample recognition that despite its wide use, power is a "slippery and problematic concept" (Martin, 1971). In Steven Lukes's influential approach to power, following Gallie (1955), power is seen as an "essentially contested concept," one of those notions which "inevitably involve endless disputes about their proper uses on the part of their users." In addition, as Lukes argues, to engage in conceptual disputes on power is itself to engage in politics (Lukes, 2004). Recently, and when addressing power in social theory, it has been recognized that there is a fundamental relation between human communication and social power, and in one of the most ambitious recent attempts to see power from a communicative angle, Manuel Castells argues as follows: "Power is primarily exercised by the construction of meaning in the human mind through processes of communication enacted in global/local multimedia networks of mass communication, including mass self-communication" (2013, p. 416).

Since the use of communicative strategies has become a pervasive process identifiable in today's environmental conflicts, struggles, and fundamental sustainability challenges, Castells's overarching statement concerning communication in the exercise of power provides an important starting point for thinking more deeply about environmental communication. Widely and descriptively speaking, environmental communication research has often been conceived as a field of studies concerned with the role that intersubjective communication plays in connection with ecosystems and the use of ecological resources (Alarcón, 2015). As a field of study, environmental communication is characterized by a focus on interpreting conflicts and environmental crises associated with political processes linked to the use of resources and human interactions with ecosystems. An important claim in environmental communication research is that it explores the constitutive dimension of communication and aims at advancing nuanced explanations of relationships between the symbolic and the material (Schwarze, 2007). Within this context, a premise that is often shared in this field of study is that the social process of communicating about ecosystems implies different assumptions, values, and beliefs about ecosystems and possible uses of ecosystem resources. Thus, Milstein (2009) conceives environmental communication scholarship as the study of how people communicate about the natural world, ascribing to environmental communication scholars the belief "that such communication has far-reaching effects at a time of largely human-caused environmental crises" (Milstein, 2009, p. 344). In turn, Robert Cox's influential work in the field in the North American context focuses on public spheres to look at the articulation of different views and meanings concerning the environment which, in his view, are articulated through environmental communication. For Cox, environmental communication should be understood as the "pragmatic and constitutive vehicle for our understanding of the environment as well as our relationships to the natural world; the symbolic medium that we use in constructing environmental problems and in negotiating society's different responses to them" (Cox, 2010, p. 37).

Issues concerning public participation in environmental management are an area of especial concern for environmental communication research (Senecah, 2004), and this is also one of the areas where a body of literature has specifically taken an environmental communication perspective to address issues of power in environmental conflicts. Todd Norton (2007), for example, employs Giddens's structuration theory to highlight what he terms the dialectic of control, where, in his view, even in contexts of great disparities in power, rarely does one agent completely monopolize power. Thus, for Norton, all agents "have some degree of power or the capacity to influence," and in his view "agents utilize power at their disposal to change circumstances and processes consistent with their desired interests." This relational view of power still needs to be more deeply connected to the existence of different interests and how those interests intersect with the deployment of environmental communication by different actors. A way to address this has been to think about power relations along with contingent relations of authority and influence. Robert Cox (2010), for instance, examines strategic dimensions of climate change communication by stating that meanings are produced within networks where power is contingent upon certain sites and through which "lines of authority and influence flow." Yet, I would argue that to separate meaning and power may analytically obscure the understanding of how and why the very acts of producing different meanings are in fact themselves acts of power. In this regard, Deborah Cox's analysis of public participation models highlights the fact that power both enables and constrains participation (Cox Callister, 2013).

These examples, taken selectively from the environmental communication literature focusing on power and communication, show us that environmental communication gets deeply entangled in the many facets of contemporary power relations in social-ecological contexts. They also suggest that it is important to more deeply analyze what I term here processes of communicative struggles, which, I argue, requires a conceptualization of communication that can be theoretically articulated with other key communicative processes such as hegemony, ideology, and discourse. In elaborating on that, I would like to briefly recall the theoretical process through which Habermas proposed his still influential theory of communicative action. For this, we need to keep in mind that Habermas first theorized about what he identified as systematically distorted communication to more specifically understand how in capitalist and class societies communication is entangled in contexts where "irreconcilable interests are recognized as antagonistic interests" (1973, p. 27). Yet, as Hanno Hardt reminds us, when thinking about communication, it is important to consider the etymological roots of the term *communication*. Hardt further highlights that communication has been applied (as a noun) to a wide variety of practices that establish, above all, commonality (2008). Thus, in following this basic conceptual dimension in our understanding of communication, one can state that considering communication as the process of "making common" is key when outlining a critical approach to communication in environmental communication research. This also means that it is important to better explain why and how the distortion of the act of making common arises, which, as Habermas suggests, originates in the confrontation of different and incompatible subjects' claims and interests in class societies where the consciousness of that incompatibility makes conflicts manifest. Within this context, what Habermas once analyzed in terms of systematically distorted communication continues to be relevant when trying to normatively assert communication as the process of "making common" in relation to contemporary social-ecological conflicts where environmental communication plays a defining role. In this regard, the argument developed by Latin American political ecologist Enrique Leff in his book *Ecology and Capital* is illuminating. For Leff, the ideological formations "covering the environmental field produce discursive practices and their function is to make neutral in the conscious of subjects the conflicts originated in divergent interests" (Leff, 1994, p. 78, trans. Cristian Alarcón). These insights, I would argue, call for a deeper exploration of the relations between power and conflicts in environmental communication to shed light on the relations between ideology, discourse, and hegemony. I argue that this opens a conceptual terrain for a more critical conceptualization of environmental communication for which the anthropology of power and the epistemologies of the South offer important insights. Thus, I shall now provide such perspectives by focusing on the paths for theorizing power and communication outlined by Eric Wolf and Paulo Freire.

Anthropologies of Power, Epistemologies of the South, and Their Relevance for a Critical and Situated Conceptualization of Environmental Communication

The task of theorizing power in social-ecological contexts characterized by conflicts and struggles for the access and use of resources links the previous discussion to ongoing conceptual discussions within anthropology (Franquesa, 2019). Within this context, authors rethinking the anthropology of power have put into question state-centric views on power since, as it is argued, state centrism obscures the multilevel processes that are inherent and contingent to power relations (Cheater, 1999).

In advancing a contemporary research agenda for the anthropology of power, Mexican anthropologist Jose Luis Escalona (2016) proposes a framework which includes the following four dimensions of a contemporary anthropology of power. First, an epistemological shift from a dualistic view of worldwide society toward a focus on ongoing, interconnected, and emigrant-oriented humankind. Second, recognition that power relationships also imply constant negotiation and struggle, that politics are a means of disputing and transforming society. Third, to continue reconsidering power, politics, and the state theoretically, and to understand power relations as differential capacities and strategies to make society, in a range of mutually constituting scales and contexts. Fourth, in addition to pure civil society, public spaces, and transparent political debates, we need to attend more deeply to realms colonized by other idioms/performances that include sorcery, witchcraft, gossip, rumor, ritual, and demagoguery (Escalona, 2016). I find it especially relevant to highlight here that when Escalona argues for a recognition of politics and power as a process of constant negotiations and struggles, and politics as a means of disputing and transforming society, he calls attention to "ways in which naming (one of the main instruments of social power) contributes to building and destroying institutions, by conceptualizing, pretending, silencing, resisting, and criticizing." Here, naming is interconnected to power in ways that imply considering power arenas of disputed domination and control, and thus imply "intentional lies and miscommunication" (Escalona, 2016). This insight is particularly relevant when analyzing the reality of political processes in environmental conflicts and everyday processes of political contestation in communicative terms. Within this context, two of

Eric Wolf's insights addressing the relations between power and communication are of particular relevance. The first is Wolf's distinction between four modes of power (2001), which, drawing also from Irene Portis-Winner (2006), I summarize as follows:

1. Power as an attribute of a person, his or her individual potency, or capability. For Wolf, this refers to the "endowment of persons in the play of power, but tells us little about the form and direction of that play" (Wolf, 2001, p. 384).
2. "Power as the ability of an ego to impose its will on an alter, in social action, in interpersonal relations. This draws attention to the sequences of interactions and transactions among people, but it does not address the nature of the arena in which the interactions transpire" (Wolf, 2001).
3. Tactical or organizational power means controls of the settings in which people may put forth their potentialities and interact with others. This takes Adams's definition of power not in interpersonal terms, "but as the control that one actor or 'operating unit' (his term) exercises over energy flows that constitute part of the environment of another actor" (Wolf, 2001; see also Adams, 1975).
4. Structural power is "power that not only operates within settings or domains but that also organizes and orchestrates the settings themselves, and that specifies the distribution and direction of energy flows" (Wolf, 2001). Here Wolf links this mode of power to what Marx identified as "the power of capital to harness and allocate labor power," and in Wolf's approach "forms the background of Michel Foucault's notion of power as the ability 'to structure the possible field of action of others'" (2001). This term rephrases the older notion of "the social relations of production" and is intended to emphasize the power to deploy and allocate social labor. For Wolf, structural power "shapes the social field of action so as to render some kinds of behavior possible, while making others less possible or impossible" (2001).

I see Wolf's modes of power as a relational approach, one which cautions us not to conceive power as a mere capability. Especially important here is how Wolf connects structural power to the social field of action. In this regard, the second important set of theoretical insights in Wolf's anthropological theory that I want to highlight has to do with how Wolf conceives power in communicative processes within conflict-laden contexts. In this regard, Wolf challenges views on communication that are oblivious to the deep connections between power, communication, and conflicts, stating, "We do not attack reality only with tools and teeth; we also grasp it with the forceps of the mind—and we do so socially, in social interaction and cultural communication with our fellows and enemies" (Wolf, 2001, p. 315). Wolf goes further and theorizes the communicative nature of conflicts in relation to the process of signification, and in criticizing some versions of symbolic anthropology, he affirms that "although signification is intimately tied up with issues of social power, 'normal' linguistics and symbolic anthropology have operated with a model of equal and power-neutral communicators or culture bearers, homogeneous speakers of language A or carriers of culture B" (2001, p. 377). In the same context, Wolf stresses that "in communication signifiers and signifieds are soldered together and supposedly arbitrary connections are in fact socially anchored and motivated" (2001, p. 378). Crucially, Wolf emphasizes here that power is deeply ingrained in the everyday use of human language: "All speakers in a linguistic community may use language, but what some people say and make others say is more fraught with social power than are the words of others" (2001, p. 378). In this regard, Wolf adds that more power-laden messages or utterances should be called "ideology," and that it is a research task "to look for these power-laden chains of signification and to study their implication for the maintenance or dissolution of social relationships" (2001, pp. 378–379).

At this point, I would like to stress that Wolf's theorizing of connections between communication and power offers a productive way to engage the anthropology of power with a critical conceptualization of environmental communication. Especially relevant here is to discuss the tendency in environmental communication research to attribute normative dimensions to communication on the environment without recognizing that conflicts and power are not an externality to different communicative practices on the environment, but are rather a constitutive aspect of such communication. In fact, if power relations are so ingrained in the communicative production and reproduction of current capitalist societies as Wolf and other have maintained, we need to first explain the role of communication in conflicts, and only after that may we be able to properly analyze how different forms of environmental communication could contribute to fostering normative alternatives to those conflicts.

These anthropological insights on power and communication help to advance a power- and conflict-oriented perspective on environmental communication. They also bring into this discussion the task of defining specific ways and contexts for environmental communication where power, knowledge, and environmental communication get entangled through conflicts. Thus, in what follows, I will continue advancing a critical conceptualization of environmental communication by bringing insights from the epistemologies of the South and the work of Paulo Freire.

I will start by observing that there is a fundamental anthropological approach in Freire's work, and this concerns precisely power and communication. As McKenna asserts,

> Freire was, in fact, an anthropological educator. He founded an educational movement based, in part, on conducting an ethnographic evaluation of a community to identify the generative themes (or 'dangerous words') which matter profoundly to people and which, for just this reason, contain their own catalytic power. (McKenna, 2013)

Bearing that in mind, I will elaborate on Freire's contemporary relevance to the theorizing of communicative struggles and to the understanding of environmental communication thereby. For this, I will focus specifically on Freire's essay "Extension or Communication" (1973) and also draw insights from his *Pedagogy of the Oppressed* (2000) to build on Freire's theorizing on communication as follows. First, Freire's works offer both a critique of existing communication and normative views on communication which are centered on conflicts and power relations in the context of struggles for resources and meaning. Second, Freire's conceptual reconstruction of the communicative dynamics in social struggles in South America opens a productive way to more deeply explore communicative struggles in terms of struggles over the very process of knowledge production. This is exemplarily shown in his critique of agricultural extension and his arguments for intercommunication in the context of his pedagogical work among peasants and extensionists during the struggles for the Agrarian Reform in 1960s Chile (Freire, 1973). Here, Freire approached micro-level processes of communication as processes of intersubjectivity in the wider context of political struggles, and he linked this to ideological and class relations of power. Third, Freire provides a dialectical understanding of communication which allows us to distinguish in clearer terms issues concerning for whom and for what communication serves, and it also provides context-specificity to communication, intercommunication, and intersubjectivity as a primary focus of attention. Fourth, Freire's analysis of social struggles develops a unique combined critique of existing communication along with the search for utopian possibilities arising from the intersubjective process of formation of sociality through communication.

Taking the previous insights together, I would argue that Freire's theorizing of communication calls attention to what I understand as spaces of communicative struggle. Here, Freire's work is lent continued relevance by a contemporary world deeply defined by social-ecological crises and struggles; ours can be seen as a time of specific communicative struggles where the materiality of expanding social-ecological crises moves along with subjective and intersubjective meaning-making within such communicative struggles. Conceptually, this means that it is not possible to simply conceive environmental communication as a neutral process. Thus, thinking about environmental communication from that perspective implies recognizing that environmental communication is produced through confrontation between different views of society and ecology, and this expresses conflictivity and struggles between mutually antagonistic sectors and classes in society. In such contexts, it is not possible to conceive environmental communication as a conflict-free process or a process that can neutrally intervene in struggles without becoming part of such struggles. Thus, bearing in mind this broad contemporary problematic in a global time and space of communicative struggles of a social-ecological nature, Freire's unique insights from situated communicative struggles offer conceptual paths to better address issues of power, conflicts, and communication concerning the following conceptual problems.

First, we can observe that, in today's environmental conflicts, the interests of mutually divergent and antagonistic sectors of society are expressed in terms of environmental communication. Thus any understanding of environmental communication in such contexts should be oriented toward theoretically explaining actors and interests in ways that fully consider conflictive social-ecological relations where environmental communication is part of antagonistic and ideological views on what societies and ecological relations are and should be. Second, and in the more particular context of analyzing the social-ecological nature of struggles and conflicts, it is important to analyze how environmental communication and discourses have in fact become an articulating process shaping struggles and conflicts in contemporary societies. Here, ideological dimensions of communicative processes are inseparable from environmental communication because environmental communication is a basic link constituting ideological relations to the environment. One way to analytically approach such conflicts is to conceive of the existence of dialectical relations between ideology and environmental communication. In turn, the combination of environmental communication and ideological processes of knowledge production give basis to environmental discourses.

Third, it is important to consider how the analysis of the processes outlined earlier and their interaction in and within the materiality of social-ecological relations may allow us to better understand hegemonic processes today. Here the production of hegemony is always the result of the articulation of communication, ideology, and discourses. In more particular terms, environmental communication articulates those three communicative processes, where hegemony should always be understood as a process and never as a static reality. In this regard, environmental communication in the context of communicative struggles for resources becomes a primary process in the production of knowledge.

The preceding elaboration, I argue, provides a conceptual background for an exploration of how a power- and conflict-oriented perspective on environmental communication helps to more deeply explore interdependencies between knowledge and power. This also entails fully considering the epistemological dimensions of conceptualizing environmental communication. The terms that capture this epistemological dimension of environmental communication are, to me, connected to what Boaventura de Sousa Santos and others have understood as epistemologies of the South (de Sousa Santos, 2012, 2015, 2018). In fact, as de Sousa Santos recognizes, one of the main sources of the epistemologies of the South is the work of Paulo Freire. For de Sousa Santos, a fundamental definitional moment of the epistemologies of the South is that these are not confined to institutional practices:

> They combine institutional and extrainstitutional practices. They are political to the extent that they constitute ways of knowing and validating knowledge that aim to contribute to the refoundation of insurgent policies capable of efficiently confronting the current, insidious, and techno-savage articulations between capitalism, colonialism, and patriarchy. Such policies, just like the epistemologies grounding them, occur inside and outside of institutions, in parliaments, governments, and judicial systems, as well as, whether

> formally or informally, in streets, squares, communities, and social networks. (de Sousa Santos, 2018, p. 248)

Thus, epistemologies of the South cannot be separated from the centrality of the struggles where institutional and extrainstitutional practices take the forms of "criticism and possibility, nonconformity and resistance, denunciation and counterproposal," which in de Sousa Santos's view can be "more or less consolidated, more or less formalized, and of longer or shorter duration" (2018, p. 248). In the context of discussing environmental issues, the central role of communicative struggles entails radical conceptual ruptures from hegemonic ways through which environmental issues are often framed and analyzed. That is the case of discourses on adaptation, resilience, sustainable development, and the like, which are oblivious to conflict-laden realities and struggles for resources. Bearing this in mind, I offer further an analysis of struggles for water justice in Chile through the lens of a power- and conflict-oriented critical conceptualization of environmental communication.

Environmental Communication and Power in the Struggles for Water Justice in Chile

An increasing number of water conflicts have surfaced in Chile during recent years in the context of a water crisis caused partly by a long drought which started in 2010. Yet, the water crisis is not only a process pertaining to lack of rain. This is a water crisis that cannot be understood without considering a legal and constitutional framework for water management that ensures private property rights over water resources in an economic context of production for export that requires enormous quantities of water. In fact, privatization of water is deeply connected with the imposition of neoliberalism in Chile (Budds, 2013). Within this context, the first assessment of water scenarios for Chile in 2019 confirmed the existence of serious water stress in the whole country, with urban and rural areas already severely hit by water scarcity. In a 2019 interview, staff assessing the water scenarios explained that

> [a]ll the [economic] sectors in the territories projected growth, and when you reviewed their long-term visions, you saw that the mining sector, the agricultural sector, the tourism sector, the sanitation sector, all of them projected growth, and you looked at the data and you said, 'Well, and with what

are they going to grow if the water [availability] decreases?' (Interview, Staff at Escenarios Hídricos, October 2019, trans. Cristian Alarcón)

Recently, the urgency of thinking and implementing policies to address water scarcity due to climate change, overconsumption, and droughts, and the increasing political opposition to the neoliberal order in Chile has created new political conditions for questioning the privatization of water resources, and the demand for ending privatization of water figured prominently during the popular uprising in October 2019. The photos in Fig. 1, taken in December 2019 in Santiago, show spray-painted political

Fig. 1 Above, the Aconcagua River completely dried up in December 2019 and stones spelling out the message: "This is not a drought; this is water pillage." Below, political messages on the walls of Santiago in December 2019: on the left, a message against forestry corporations, and on the right, a message against water privatization and demanding the right to water. (Photos: Cristian Alarcón Ferrari)

messages focused on water and environmental conflicts during the protests in 2019 and stone-made messages in a dry riverbed in the Valparaíso Region of central Chile.

In an attempt to deal with water scarcity, during the summer of 2020 the government launched a campaign to reduce water consumption which rested on private consumption decisions, particularly in Santiago, but without addressing the structural problems concerning water management and inequality in the current water crisis. Yet recently, a legal reform discussed in the parliament considered giving the national government the power to ensure and give priority to human consumption of water in cases of serious water stress and scarcity. On the other hand, the government launched a water roundtable to address the water crisis. Though they envisioned a roundtable with the limited participation of selected parties and the clear objective of keeping private property of water untouched, the new political context after the popular uprising in October 2019 forced the government to facilitate open discussions on the water crisis and strategies to face it. Thus, the water roundtable was open to public consultations and became a space for communicative struggles focused on the current legislation regulating the access to water resources (personal observation during an online seminar of the water roundtable in October 2020). This can also be seen in the results of the nationwide surveys organized by the government to assess opinions on water management, where 77.6% of respondents agreed that a basic principle in water legislation should be to "ensure continuous and sufficient access to water as a fundamental human right for personal and domestic use" (Mesa Nacional del Agua, 2020).

This needs to be understood in relation to the long struggle of local movements for water justice in Chile, and the resistance and alternatives articulated by these different territorial movements in the middle of the local social-ecological conflicts caused by the water crisis. The communication and material practice of these movements can be understood through the lens of claims to water justice, which raises important prospects for bottom-up meanings of water democracy. Water justice has been conceived as a basic demand for protection of water resources for human consumption and as a demand to ensure the sustainability of water resources at the local levels where water is extracted, but often for consumption in other places (Boelens et al., 2018). In turn, water democracy can be understood here as a situation where water users struggle and define strategies to participate in water decision-making and "do not passively bow to the top-down imposition of water policies and governance modes" (Hoogesteger, 2017). In what follows, I will show how environmental communication, power, and conflicts are constitutive aspects in the quest for water justice and water democracy in Chile.

First of all, there are three factors that are key to consider when analyzing and explaining how and why struggles for water justice and democracy have become so important in Chile recently. First, movements for water justice have constantly criticized the system of private property rights to water established through the privatization process under the Pinochet military dictatorship and maintained, and even deepened, during recent elected governments. Second, we have the social-ecological inequalities of the capitalist development of Chile's export-oriented economy, which is still based on exports originating in four economic activities that demand enormous quantities of water: mining, forestry and pulp production, fisheries, and agribusiness. Third, there is a megadrought that has lasted 10 years now and has aggravated water scarcity. In some areas this drought has coincided with the expansion of avocado plantations, which are now in places where extraction and accumulation of water make irrigation systems possible in dry areas (Duran-Llacer et al., 2020), and eucalyptus plantations, which are also associated with high water demand in dry areas (Alvarez-Garreton et al., 2019). Though eucalyptus plantations are not irrigated, their effects in some rural areas of southern Chile are similar to those of the avocado plantations in the north. In some other rural areas, the drought aggravates conflicts over water arising from projects to dam rivers for hydroelectrical and irrigation objectives. In what follows, I more deeply analyze such water conflicts by focusing on the three different regions, shown in Fig. 2, where I have carried out fieldwork and where the aforementioned factors interplay in water conflicts.

In the Santiago Metropolitan Region, where Chile's capital, Santiago, is located, about 70% of the drinking water is obtained from rural areas crossed by the Maipo River, approximately 35 km from downtown Santiago. Drinking water obtained from the Maipo River is commercialized by one large company, which needs to purify the water before delivering it to consumers in Santiago. This large private water company, Aguas Andinas, owned by the Spanish international water company Suez, was previously a publicly owned company, but its privatization began during the dictatorship, and it was then fully privatized in the post-dictatorship period. The company depends on water availability in the Maipo River and

Fig. 2 Locations of fieldwork in the Valparaíso, Santiago, and Ñuble regions of Chile. (This image used with permission of B1mbo, CC BY-SA 3.0 CL, via Wikimedia Commons)

on the Yeso Dam in the Andes foothills. Due to the lack of snow and rain in 2019, the dam had only 37% of its capacity in January 2020, raising serious concerns about water availability in the future. In March 2020, rural areas in the Maipo Basin were declared to be in a state of water scarcity, and though in 2020 rains ameliorated the effects of the drought, this was not enough to end the drought. Due to water scarcity in the Maipo Valley, the water company is making investments to access more water sources and ensure water availability for Santiago, and its projects include well drillings in the suburbs of the city. In parallel with issues concerning water scarcity, the Maipo River is the site of a long conflict between local rural communities and national environmental groups that are opposing the Alto Maipo Hydroelectric Project. This project consists of damming and redirecting the Maipo River to produce and transfer electricity for consumption in other parts of the country, including mining industries in the north. A point of contestation here is the fact that the Alto Maipo Hydroelectric Project is based on a private commercial agreement between the water company and the Alto Maipo hydroelectric company. This private agreement ensures the hydroelectrical company the right to use water resources for which the water company has user rights. Thus, water use in these rural areas is mainly regulated as a private agreement between these two private companies. The threats to water resources implied by this project are not the only ones for the population of the city of Santiago. Recently, the same water company raised concerns about a mining project planned in another valley providing water to the city. Here, a main concern is that the Environmental Impact Assessment (EIA) for this mining project has not fully taken into account how it may affect water resources. Taking these two examples together shows that approaching the water crisis in Santiago requires an understating of the legislation contained in a water code that ensures private property rights to water resources. This water code is a product of the neoliberal project imposed in the country during the dictatorship, and reforms to change the property regime established in this water code proposed by several coalitions, including social-environmental movements, academics, and political movements, are resisted by business associations and politicians aligned with the interests of such businesses. Thus, water provision in Santiago is characterized by an institutional arrangement originating in the privatization of the main water company, the establishment of private property rights to water (which benefits the private water company, other economic organizations, and private owners), and the lack of public control on how to manage water resources. In this regard, Chilean water legislation is known as one of the most neoliberal water regulation policies in the world, as it ensures private property rights to water resources, and markets and private actors can act in the water markets with few public regulations (Bauer, 2015). In addition, privatization of water is deeply linked to inequal distribution of the now scarce water resources in the city of Santiago (Durán, 2015).

The water crisis in Chile has also dramatically materialized in the province of Petorca in the Valparaíso Region. While the drivers of the water crisis here are several, including less rain and snow in the Andes associated with the drought and climate change, the water conflicts are basically rooted in legal and illegal appropriation of water resources for agribusinesses in the context of water scarcity. In this regard, conflicts are associated with the aggressive expansion of avocado and citrus plantations for export, which dramatically reduces water availability for human consumption in the area. In this case, plantation owners use their private property rights to water to ensure irrigation and production, and in some cases, there has even been illegal appropriation of groundwaters to irrigate the plantations (Bolados García et al., 2018). As the pictures in Fig. 3 show, and as explained by informants during personal observations in the area, these avocado plantations are only possible because of the existence of water accumulators, which source water in dry areas where no plantations existed until recently. These accumulators obtain water through

Fig. 3 Above, avocado plantations in the Andes foothills close to Petorca in the Valparaíso Region of Chile. Below, water accumulators for irrigation of avocado plantations and a dried-up stream close to the water accumulators. (Photo: Cristian Alarcón Ferrari)

pipelines that connect different water extraction points in the Andes foothills and have been opposed by local movements and activists, who organize their demands around defending water resources for the local communities and ending the privatization of water and the system of private property rights to water resources. Crucially, they also demand support for alternative ways to access and use water resources, with a focus on human consumption as well as more sustainable food production systems.

Another region where the water crisis is especially noticeable is the Ñuble Region in southern Chile, where industrial forestry plantations of exotic tree species such as eucalyptus and pines along with increasing irrigations needs for agribusinesses have led to local conflicts over water resources. In the case of industrial tree planting, increasing evidence of water scarcity due to fast-growing and water-intensive tree plantations have mobilized local actors to find ways to regulate and even stop tree planting. With regard to dam projects, local communities have opposed a major dam project which will also affect important native forest resources. Here, opposition during the EIA process did not succeed, and the company eventually obtained an environmental authorization to build the dam. Also, a hydroelectrical component project which was added to the project only recently obtained an environmental authorization through an EIA. Recently, hundreds of activists were able to stop the public auction for the right to use water from another major river and a watercourse in the region. These auctions are allowed in the Chilean legal system and, if successful, they entail exclusive rights to water. In these cases, demands for water use and water rights articulate political resistance to the privatization of water, but in some cases resistance also leads to land-use alternatives. This second form of resistance is represented by agroecological movements in the region, which have confronted the water crisis by elaborating and implementing agroecological systems that allow less water-intensive agriculture and the use of species that consume less water (Interviews and observations in the Ñuble Region, December 2019).

When these water conflicts are taken together, the material and communicative articulation of popular movements defending water resources and demanding rights to water use reveal themselves to be a distinctive aspect of the struggles for water justice in Chile. Two characteristics of these movements can be considered in greater detail here. First, these are movements that communicatively question the legitimacy of the system of private property rights to water. Thus, at the center of this dispute are the political meanings of property rights to water resources and how they are contested. Second, these movements aim at giving a material meaning to demands for ensuring water use as a common good and the priority of human consumption of water. I would argue that this provides important grounds for empirically deepening a critical analysis of environmental communication in these water conflicts.

A Critical Environmental Communication Analysis of Water Conflicts in Chile: Knowledge and Power over Water and Epistemologies of the South

The struggles for water justice in Chile are not only multidimensional in the material sense but also multidimensional at the level of theory and epistemology. Thus, the analysis of environmental communication in this context needs to encompass the process of knowledge production and communicative struggles implied in giving meanings to rights and uses of water. In this sense, one can discern in the struggles for water justice in Chile that when local movements aim at protecting water resources for human consumption, they are at the same time struggling to create new meanings for water decision-making. Thus, in analyzing and explaining these struggles by deploying a critical conceptualization of environmental communication, the following prominent relations can be established. First, these water conflicts show how spaces of communicative struggles concerning water are brought about by the interests of different actors trying to produce and impose meanings on the problem and on the possible solutions to the conflicts. The struggle around private property rights to water shows this clearly. On the one hand, for those that ideologically and materially defend private property rights to water, this defense is part of their struggle to maintain the hegemony of the neoliberal project over water, and they use their power to defend private property rights and market mechanisms governing water resources by accommodating this defense to the context of sustainability concerns. In fact, one of the discourses reproduced in the water debate in Chile articulates the idea that property rights are part of the solution to the water crisis, and thus private water companies and agribusinesses have discursively engaged in active campaigns to show environmental credentials and alignment with sustainable development goals. In these intersections between environmental communication and power lie some of the main opposing forces articulating water conflicts in Chile.

Second, as these cases also show, environmental communication is locally situated, conflict-oriented, and entangled in the wider conflictivity of neoliberal capitalism and the ideological and discursive articulation of hegemony in the environmental-policy-making process in Chile. On the other hand, the rise of movements for water justice in Chile shows how novel counterhegemonic forms of knowledge are today able to firmly articulate social power and so to materialize meanings and discourses of water justice in the country.

Third, these are struggles that refer constantly to the process of creating power to make water a common good, and they are articulated through the production of meaning about material access and control to water resources. This indicates that to critically theorize environmental communication in these specific contexts is an important step in understanding how and why power and environmental communication are inseparably linked within water conflicts in Chile. To theorize this, it is key to recognize the epistemological dimensions of the enactment of environmental communication by activists and organizations taking the side of the right to water and water justice in the struggles for water in Chile. In fact, a condition of possibility for these movements is the wide networks of communication and education and the processes of situated knowledge production which are communicatively expanded to form wider public spheres. As the process initiated by the water roundtable shows, the efforts to depoliticize the water discussion in Chile are impossible today because open channels of communication also mean freer ways to politicize the water discussion. Thus, demands for water democracy are the enactment of a political goal for water justice long in the making, and they enable us today to think about water justice in a much clearer political way. In a context like this, as de Sousa Santos suggests, we see the creative potential of the epistemologies of the South: If these struggles have so deeply resonated in the widespread rejection of neoliberalism that moved millions of Chileans to massively build the popular uprising of 2019 and then vote for a new constitution in 2020, it is because, to a crucial degree, another way of knowing the water crisis also enacted another way of conceiving social power over water. Thus, it is the social power of the people, their use of environmental communication versus corporate environmental communication, and the long struggles for water justice that today make the critique of property rights to water so pivotal in the political demonstrations against neoliberalism in Chile. And this is something that neither the government-led roundtable on water nor corporate environmental communication and their neoliberal-centric epistemology can accommodate or domesticate today.

Concluding Remarks

This chapter has offered some conceptual and theoretical insights to advance a conflict- and power-oriented perspective for a critical conceptualization of environmental communication. By drawing insights from Paulo Freire, Eric Wolf, the contemporary anthropology of power, and the epistemologies of the South, I have attempted to forge some conceptual links between environmental communication and anthropology. What emerges from this attempt, I argue, is one path to deepening our understanding of how and why environmental communication is ingrained in struggles over the meaning and materiality of social-ecological conflicts today. Furthermore, by establishing a clear link between this approach to environmental communication and the question of knowledge and power relations in the struggles for water justice in Chile, the chapter has brought into the discussion the role that environmental communication plays in shaping communicative and epistemological struggles. In this regard, the voices of people fighting for water justice in Chile show how environmental communication gets entangled with the everyday production of epistemologies of the South. In fact, water conflicts in Chile cannot be understood without approaching the overarching epistemological interests at stake in each local water conflict. In these contexts, as elsewhere, environmental communication is located at the center of such struggles and conflicts. To conclude, I would argue that this kind of critical theorizing of environmental communication is important today not only to better understand and explain the meanings of normative views of environmental communication but also to understand how such normative views cannot be separated from the interests of subjects situated in contingent social-ecological relations and conflicts.

Acknowledgments I am grateful for the very helpful comments, editorial support, and suggestions of Annelie Sjölander-Lindqvist and Michael Dove, and for the comments and suggestions from two anonymous reviewers. I would also like to thank the participants in the panel on Environmental Communication and Anthropology at the 2018 Applied Anthropology Conference in Philadelphia for their comments on the conference paper on which this chapter is based. This research was carried out during a postdoctoral fellowship at the Department of

Development Sociology at Cornell University, supported by a postdoctoral scholarship granted by Chile's National Commission for Scientific and Technological Research—CONICYT (74170059), and during the author's affiliation with the Law School at the Universidad Academia de Humanismo Cristiano, Chile. Also, part of the empirical work was supported by a grant from FORMAS (the Swedish Research Council for Sustainable Development, grant number 2018-02341).

References

Adams, R. N. (1975). *Energy and structure: A theory of social power*. University of Texas Press.

Alarcón, C. (2015). *Forests at the limits. Forestry, land use and climate change from political ecology and environmental communication perspectives: The case of Chile & Sweden*. Doctoral Thesis No. 2015: 5, Faculty of Natural Resources and Agricultural Sciences, Swedish University of Agricultural Sciences.

Alarcón, C. (2019). Transforming wood energy in Sweden and Chile: Climate change, environmental communication and a critical political ecology of international forestry companies. *Critical Perspectives on International Business, 16*, 361–377. https://doi.org/10.1108/cpoib-05-2018-0039

Alarcón, C. (2020). Crisis socioecológicas y educación popular ambiental en el mundo rural: la relevancia de Paulo Freire para los estudios críticos de la comunicación ambiental y la educación para el desarrollo sostenible. Paulo Freire. *Revista de Pedagogía Crítica, 24*, 149–171.

Alvarez-Garreton, C., Lara, A., Boisier, J. P., & Galleguillos, M. (2019). The impacts of native forests and forest plantations on water supply in Chile. *Forests, 10*, 473. https://doi.org/10.3390/f10060473

Bauer, C. J. (2015). Water conflicts and entrenched governance problems in Chile's market model. *Water Alternatives, 8*(2), 147–172.

Boelens, R., Vos, J., & Perreault, T. (2018). Introduction: The multiple challenges and layers of water justice struggles. In R. Boelens, T. Perreault, & J. Vos (Eds.), *Water justice* (pp. 1–32). Cambridge University Press.

Bolados García, P., Henríquez Olguín, F., Ceruti Mahn, C., & Cuevas, A. S. (2018). La eco-geo-política del agua: Una propuesta desde los territorios en las luchas por la recuperación del agua en la provincia de Petorca (Zona central de Chile). *Revista Rupturas, 8*, 159–191.

Budds, J. (2013). Water, power, and the production of neoliberalism in Chile, 1973–2005. *Environment and Planning D: Society and Space, 31*, 301–318.

Castells, M. (2013). *Communication power*. Oxford University Press.

Cheater, A. (1999). Power in the postmodern era. In A. Cheater (Ed.), *The anthropology of power* (pp. 1–12). Routledge.

Cox Callister, D. (2013). Land community participation: A new "public" participation model. *Environmental Communication: A Journal of Nature and Culture, 7*, 435–455.
Cox, R. (2010). *Environmental communication and the public sphere*. Sage.
de Sousa Santos, B. (2012). Public sphere and epistemologies of the South. *Africa Development, 37*, 43–67.
de Sousa Santos, B. (2015). *Epistemologies of the South: Justice against epistemicide*. Routledge.
de Sousa Santos, B. (2018). *The end of the cognitive empire: The coming of age of epistemologies of the South*. Duke University Press.
Durán, G. (2015). Agua y pobreza en Santiago de Chile: Morfología de la inequidad en la distribución del consumo domiciliario de agua potable. *EURE (Santiago), 41*, 225–246.
Duran-Llacer, I., Munizaga, J., Arumí, J. L., Ruybal, C., Aguayo, M., Sáez-Carrillo, K., Arriagada, L., & Rojas, O. (2020). Lessons to be learned: Groundwater depletion in Chile's Ligua and Petorca watersheds through an interdisciplinary approach. *Water, 12*, 2446.
Escalona, J. L. (2016). Anthropology of power: Beyond state-centric politics. *Anthropological Theory, 16*, 249–262.
Franquesa, J. (2019). Resources: Nature, value and time. In J. G. Carrier (Ed.), *A research agenda for economic anthropology* (pp. 74–89). Edward Elgar Publishing.
Freire, P. (1973). Extension or communication (L. Bigwood & M. Marshall, Trans.). In *Education for critical consciousness* (pp. 91–164). Continuum International Publishing.
Freire, P. (2000). *Pedagogy of the oppressed*. Bloomsbury Publishing.
Gallie, W. B. (1955). Essentially contested concepts. *Proceedings of the Aristotelian Society, 56*, 167–198.
Habermas, J. (1973). *Legitimation crisis*. Beacon Press.
Hardt, H. (2008). *Myths for the masses: An essay on mass communication*. Wiley.
Hoogesteger, J. (2017). The politics of water democracy: Insights from grassroot struggles in the Ecuadorian Highlands. *Asia Pacific Viewpoint, 58*, 74–85.
Joas, H., & Knöbl, W. (2009). *Social theory: Twenty introductory lectures*. Cambridge University Press.
Leff, E. (1994). *Ecología y capital: Racionalidad ambiental, democracia participativa y desarrollo sustentable*. Siglo XXI.
Lukes, S. (2004). *Power: A radical view*. Red Globe Press.
Martin, R. (1971). The concept of power: A critical defence. *The British Journal of Sociology, 22*, 240–256.
McKenna, B. (2013). Paulo Freire's blunt challenge to anthropology: Create a pedagogy of the oppressed for your times. *Critique of Anthropology, 33*, 447–475.

Mesa Nacional del Agua. (2020). Participación ciudadana. Primer informe consulta digital Mesa Nacional del Agua, Chile.

Milstein, T. (2009). Environmental communication theories. In S. W. Littlejohn & K. A. Foss (Eds.), *Encyclopedia of communication theory* (pp. 345–349). Sage.

Norton, T. (2007). The structuration of public participation: Organizing environmental control. *Environmental Communication, 1*, 146–170.

Portis-Winner, I. (2006). Eric Wolf: A semiotic exploration of power. *Sign Systems Studies, 34*, 339–356.

Schwarze, S. (2007). Environmental communication as a discipline of crisis. *Environmental Communication, 1*, 87–98. https://doi.org/10.1080/17524030701334326

Senecah, S. L. (2004). The trinity of voice: The role of practical theory in planning and evaluating the effectiveness of environmental participatory processes. In S. P. Depoe, J. W. Delicath, & M.-F. Aepli Elsenbeer (Eds.), *Communication and public participation in environmental decision making* (pp. 13–33). SUNY Press.

Wolf, E. R. (2001). *Pathways of power: Building an anthropology of the modern world*. University of California Press.

Commentary

Camilla Sandström and Katrina Rønningen

Introduction

In this concluding commentary, we have been asked to give an "outsider's view" through our disciplines of geography and political science of the overall objective of the book, that is, how an anthropological perspective can further our understanding of the diversity of environmental communication (see Sjölander-Lindqvist, chapter "Introduction"). Based on the eight studies in the book, we ask: To what extent do we, from our different disciplines, read or interpret the texts in similar or different ways?

As a geographer and a political scientist, we have an understanding similar to the one highlighted by editor Sjölander-Lindqvist in the book's introduction, namely that any situation where there may be divergent

C. Sandström (✉)
Political Science, Umeå University, Umeå, Sweden
e-mail: camilla.sandstrom@umu.se

K. Rønningen
Ruralis – Institute for Rural and Regional Research, Trondheim, Norway
e-mail: katrina.ronningen@ruralis.no

A. Sjölander-Lindqvist et al. (eds.), *Anthropological Perspectives on Environmental Communication*, Palgrave Studies in Anthropology of Sustainability, https://doi.org/10.1007/978-3-030-78040-1_10

understandings regarding land, places, and living beings should be approached as contingent and embedded in contexts of interacting interests, values, short- and long-term objectives, ideologies, and particular norms of those concerned.

However, these types of nested situations may, despite a similar understanding, be interpreted in quite different ways depending on the discipline and the theoretical underpinnings in focus but also on methodological choices. Social sciences have many overlapping theoretical and methodological approaches; in fact, the different disciplines are inspired by similar ideas about society developed by scholars all the way from Hegel to Habermas. Depending on the discipline and subfield, geography, for example, contributes an analytical perspective on the spatial effects of human–environment interaction. Political science, for example, includes the study of institutions, governance, politics, democratic representation, and participation. Anthropology, which underpins this book, studies human experience; this includes the exploration of the worldviews, ways of life, and forms of knowledge surrounding human environmental existence, as Sjölander-Lindqvist states in the introduction. What unites anthropology, geography, and political science is their focus on power and its structuring effects. However, this unity is challenged by the lack of a common language and common definitions of key concepts between disciplines. In retrospect, after reading the various chapters of this book, we acknowledge the need to move beyond disciplinary comfort zones in order to engage productively with the different perspectives and contributions that each field has to offer.

Environmental communication, which is a truly interdisciplinary research field, may in that sense have the potential to bridge the gap, not only between science and society but also between our different disciplines. In its extension, environmental communication may thus contribute to advancing the understanding of the interplay of the environment and political, economic, social, and cultural factors in practice, but also to advancing the ontological, epistemological, and theoretical positions of the different disciplines. We will return to this point at the end of our commentary.

One of the overarching aspects that we have identified among the various chapters of the book is what can be defined as the need to establish a dialogue between the many different "ways of seeing" the world by the different groups and actors. This requires a much more fundamental sensitivity toward contexts and cultural, historical, and socio-economic factors, including spiritual factors. It is fundamentally important to prevent what Joosse et al. refer to as "discursive colonization" (2020, p. 6), that is, "the reproduction of the interests of the powerful through certain narrowly defined forms of knowledge and scholarship" (Stoffle, chapter "Cultural Transmission in Slovak Mountain Regions: Local Knowledge as Symbolic Argumentation" of this book).

We see that many of the texts in this book carry this sensitivity. We also see the present and future of the local communities and places presented through these eight cases as being challenged by various types of "modernization processes" coming from a profit-seeking perspective that don't seem to leave the local communities better off. Social change, power, and identity are at the core of these processes, and the ongoing various ways of trying to handle environmental communication. In many of the cases, environment, climate change, and adaptation are the emperor's new clothes for exercising power and control over environmental resources, landscapes, and ultimately profit making.

Environmental communication is also exercise of power. The many mandatory requirements of planning and licensing processes, often ambitious in terms of participatory processes, local involvement, and so on, are good examples of *how* environmental communication is or may be carried out, as part of exercising power. For example, while Environmental and Social Impact Assessments (EIAs) are required in many countries as a tool in the planning processes of larger environmental and landscape interventions, they too often fail to fulfill their function, or become part of further power and communicative structures, creating conflicts and further injustice as a result (Stoffle & Minnis, 2008; Stoffle et al., 2013). Also, the cheapest bids for carrying out the EIAs often win, and the quality of some of these processes may be questioned. Furthermore, these planning processes and legal requirements in many cases do not include social impacts, which in turn may lead to power struggles and the mobilization of people, actions, and resources through, and as a part of, environmental communication (Eckerd, 2017). In the following part we will, based on our reading of the chapters, highlight four different aspects of these power struggles and mobilizations of people.

Attempts to Decolonize Land While Challenging Modernization

Several of the chapters in this book highlight the role of environmental communication as a tool among Indigenous Peoples around the world in processes of decolonization and struggles for self-determination. Indigenous Peoples have, with varying degrees of success, reclaimed histories and cultures that have been ignored or misinterpreted by researchers and land managers in order to reconnect to traditional lands. However, as highlighted by Van Vlack (chapter "Dancing with Lava: Indigenous Interactions with an Active Volcano in Arizona"), in other places it may be an ongoing struggle which, as part of decolonization processes, also tends to question modernization in terms of its ethnocentric or even Eurocentric worldview.

The empirical case of the Southern Paiute people near the Little Springs Lava Flow in northern Arizona, USA, is illustrative as a case where there exist differences in interpretation of the landscape and its uses over time between Indigenous People on the one hand, and researchers and land managers on the other. While it may be difficult to bridge between different epistemologies in this specific case, Van Vlack suggests that, instead of continuing to exclude the Southern Paiute people from the governance and management of the area, one solution could be to open these processes up to multiple voices. Environmental communication could then be used to further explore the landscape and its use while at the same time promoting social learning.

The case study of the Iliamna Lake Central Yup'ik Place Name Project in Southwest Alaska in the chapter "Community Voices, Practices, and Memories in Environmental Communication: Iliamna Lake Yup'ik Place Names, Alaska" by Kugo is an illustrative example of this type of learning process, where the recognition of indigenous place names not only empowers the Indigenous population but it may also contribute to improved relationships between Indigenous Peoples and, for example, authorities in terms of communication needs.

Despite the many good examples in the book, the chapters also confirm that more or less mandatory tools for assessing consequences of new land utilization, such as EIAs, generally lack the tools to take multi-generational experiences, oral narratives, and local knowledge into consideration. The planning tools often rely on a specific type of knowledge and use very narrow time perspectives. As the EIAs often are carried out by competing consultancy companies, the cheapest bid often wins. In one recent example from wind energy development, two separate areas in Norway had their EIAs carried out in a total of three days according to their own records. These investigations "are so insufficient, they are almost without value," a county governor representative stated (Thunold et al., 2021, trans. Katrina Rønningen). In the chapter "Living Stone Bridges: Epistemological Divides in Heritage Environmental Communication," Stoffle demonstrates how understandings of landscape are based in long-term interaction with the environment, which is a feature of communication that is also discussed by Sjölander-Lindqvist in the chapter "Arsenic Fields: Community Understandings of Risk, Place, and Landscape."

In other words, very short-term observations may be used as a basis for decisions with profound consequences. These processes, which are sometimes designated as "neo-colonialism" or "green colonialism," largely fail to take different ways of knowing into consideration. Anthropological methods on the other hand, often based on long-term field work, may through their methodology provide an alternative approach that is sorely needed.

Hence, environmental communication could be one way of including and integrating local and indigenous knowledge. However, this also requires a more thought-through planning process, more funding, and another type of competence, namely anthropology, to be able to map land use properly. Another approach would be to insist that new activities in an area need to operate according to seasonal movements. Rhythm analysis, for example, may allow for multifunctional and multipurpose use, and a way to integrate many various needs, wishes, and purposes (Flemsæter et al., 2019). The exploration of place names may reveal environmental and spatial information, but also emphasize the temporal and spiritual relationships between the people and the land (Kugo, chapter "Community Voices, Practices, and Memories in Environmental Communication: Iliamna Lake Yup'ik Place Names, Alaska").

Acknowledgment of a Diversity of Thoughts

While several global assessments such as IPBES (2018, 2019) and the Global Environmental Outlook (2019) have opened themselves up to the idea that we live on one planet, but in multiple worlds, the society–nature dichotomy is still the prevailing way in which states and societies are organized (Sjölander-Lindqvist et al., 2020). Indigenous worldviews and different understandings and conceptions of nature, as well as society–nature relations are rarely acknowledged in practice, and when they are, it is often an add-on to existing environmental policy, programs, or projects, instead of being in epistemological parity with them.

The case of Living Stone Bridges by Stoffle (chapter "Living Stone Bridges: Epistemological Divides in Heritage Environmental Communication") illustrates this very well in what is defined as discursive colonization having an effect both on what type of knowledge, and thus also who is recognized and involved, and on what grounds, in the governance and management of protected areas.

The same pattern is repeated all over the world, including cases where Traditional Ecological Knowledge (TEK) is recognized, such as at the Laponia World Heritage site in Sweden. TEK is acknowledged, but not on its own terms; instead it is for the sake of biodiversity (Reimerson, 2015), for example. Since indigenous conceptions of nature vary, as each ethnic group has their own way of envisioning nature and understanding the relations that come with it, an appreciation of TEK also requires an appreciation of diversity in thought, worldviews, and values (Berkes, 2012).

Environmental communication could play an important role as a bridge between different worldviews—both between scientific disciplines and in particular as a kind of mediator between different ways of knowing. Furthermore, while acknowledging TEK, which often is holistic and not constructed on the basis of the society–nature dichotomy and other modern dichotomies like body and spirit, we may be able to—in collaboration—develop what is often called for: more encompassing and holistic views on the governance and management of the environment.

Exploitation and Commodification of Natural Resources and Knowledge

Developing Traditional Ecological Knowledge (TEK), local, indigenous knowledge and language into a "business model" that local and indigenous groups can monetize is one optimistic approach. Crucial knowledge of many types of land use and management could be gathered into maps, GISs, and GPSs. This could potentially increase these communities' social capital, making them and their knowledge relevant through a modernizing process. But if that happens, then crucial cultural heritage is commodified, ushering in all the problems associated with that. Still, what is the alternative? Will purist approaches keep them irrelevant?

There is potential for both conserving and using indigenous and local knowledge by commodifying it. It may give this knowledge a function in contemporary society. As demonstrated by Murin (chapter "Cultural Transmission in Slovak Mountain Regions: Local Knowledge as Symbolic Argumentation"), this is dependent on TEK being shared across generations. This transmission of knowledge is challenged by modernization processes, as we can see from Murin's case study of remote mountainside settlements in Central Slovakia, where the change and abandonment of traditional land-use practices, due to outmigration to urban centers and aging rural populations, has implications for the ability of the local community to manage community-based agricultural resources and protect the cultural landscape.

However, there are numerous examples where the commodification of indigenous knowledge, plants, and other resources leaves nothing to the communities that developed or traditionally utilized and survived upon them, except for the knowledge that they have been robbed. Intellectual property rights and patent rights are part of this. The core issue here is how to both protect and monetize cultural heritage. Commodification involves a high risk of exploitation by outsiders—who owns it, and who has the right to exploit it?

The MBT Maasai sandal is one such example. As a reaction, the Maasai Intellectual Property Initiative Trust educates the community about the value of their brand and hires lawyers "to persuade multinational companies to recognize the Maasai trademark—and pay for it" (Pilling, 2018). Intellectual property rights, patents, and so on comprise a huge and difficult industry in themselves, dealing with which requires resources that many small (indigenous) communities do not have.

Another example of what are termed processes of communicative struggles is the chapter "Power, Conflicts, and Environmental Communication in the Struggles for Water Justice in Rural Chile: Insights from the Epistemologies of the South and the Anthropology of Power" by Alarcón. Questioning the very nature of property rights and to what extent it is possible or even morally right to exploit common goods such as water, Alarcón shows how environmental communication becomes entangled with everyday production of epistemologies and thus cannot be understood without taking this into consideration. Environmental communication may thus be used to understand power struggles and conflicts, but it may also be understood as shaping these struggles and as such be a double-edged sword.

Risk Communication, Perception, and Agency

Risk communication is a crucial part of environmental communication. The story told by Sjölander-Lindqvist in the chapter "Arsenic Fields: Community Understandings of Risk, Place, and Landscape," describing arsenic fields left by copper mining in Sweden, how local residents negotiate the meaning of place, and how identity and loyalty are important to the place, is in many ways touching. At the same time, it reminds us that, for a large part of the world's population, there is no way out. How do you then deal with information on environmental risk to very poor people with no alternatives in highly contaminated areas, which, if you take it seriously, will set severe limitations on practically all activities—from having children to eating the produce from the land. And if we take it one step further: How do we all relate to the communication on climate change that is altering the entire world?

Concluding Remarks

While reading this book from the perspectives of geography and political science, we have recognized the need to identify the many "ways of seeing" the world, which requires sensitivity toward contexts and cultural, historical, and socio-economic factors, but also acknowledging the need to include different "ways of knowing" in order to be able to admit and potentially integrate different ontologies and epistemologies. Finally, and from our specific perspectives, we would like to highlight the need to also acknowledge different "ways of doing"; in other words, how to turn the different ways of seeing and knowing into legitimate regulations, processes, institutions, and legal frameworks for the potential sharing of benefits and burdens linked to natural resources and places.

In the chapter "'The Sea Has No Boundaries': Collaboration and Communication Between Actors in Coastal Planning on the Swedish West Coast," Larsson and Sjölander-Lindqvist highlight this particular need in order to understand the processes of environmental communication and how they are or become embedded in institutions and guide the interaction between actors in specific institutional settings.

More specifically, ways of doing can be understood as different modes of governance and management (see also Mårald et al., 2017), where collaborative governance, as in the case of coastal planning, requires navigating a context where power is distributed across diverse societal subsystems and among many actors with different ways of knowing, ideas, and desired benefits. To be able to handle such complex socio-ecological realities, mutual interactions and learning across social levels are key. Environmental communication may, as we have seen in this book, contribute to bridging this gap by integrating different ways of knowing. Not the least, environmental communication, with its strong links to and further developed within anthropology, may also contribute to bridging the gap between the ontological, epistemological, and theoretical positions of our different disciplines by "further[ing] our understanding of [...] the different ways people—verbally and non-verbally—communicate about and with their surrounding environments" (Sjölander-Lindqvist, chapter "Introduction", p. 2).

References

Berkes, F. (2012). Implementing ecosystem-based management: Evolution or revolution? *Fish and Fisheries, 13*(4), 465–476. https://doi.org/10.1111/j.1467-2979.2011.00452.x

Eckerd, A. (2017). Citizen language and administrative response: Participation in environmental impact assessment. *Administration & Society, 49*(3), 348–373. https://doi.org/10.1177/0095399714548272

Flemsæter, F., Gundersen, V., Rønning, K., & Strand, O. (2019). The beat of the mountain: A transdisciplinary rhythm analysis of temporal landscapes. *Landscape Research, 44*(8), 937–951. https://doi.org/10.1080/01426397.2018.1535652

IPBES. (2018). The IPBES regional assessment report on biodiversity and ecosystem services for Europe and Central Asia. In M. Rounsevell, M. Fischer, A. T.-M. Rando, & A. Mader (Eds.). Secretariat of the Intergovernmental Science-Policy Platform on Biodiversity and Ecosystem Services.

IPBES. (2019). Global assessment report on biodiversity and ecosystem services of the Intergovernmental Science-Policy Platform on Biodiversity and Ecosystem Services. In E. Brondizio, J. Settele, S. Díaz, & H. Ngo (Eds.). Secretariat of the Intergovernmental Science-Policy Platform on Biodiversity and Ecosystem Services.

Joosse, S., Powell, S., Bergeå, H., Böhm, S., Calderón, C., Caselunghe, E., Fischer, A., Grubbström, A., Hallgren, L., Holmgren, S., Löf, A., Nordström Källström, H., Raitio, K., Senecah, S., Söderlund Kanarp, C., von Essen, E., Westberg, L., & Westin, M. (2020). Critical, engaged and change-oriented scholarship in environmental communication: Six methodological dilemmas to think with. *Environmental Communication, 14*(6), 758–771. https://doi.org/10.1080/17524032.2020.1725588

Mårald, E., Sandström, C., Nordin, A., & others. (2017). *Forest governance and management across time: Developing a new forest social contract*. Routledge.

Pilling, D. (2018, January 19). Warrior tribe enlists lawyers in battle for Maasai "brand". *Financial Times*.

Reimerson, E. (2015). *Nature, culture, rights: Exploring space for indigenous agency in protected area discourses*. Doctoral dissertation, Umeå universitet. Retrieved March 3, 2021, from http://urn.kb.se/resolve?urn=urn:nbn:se:umu:diva-110737

Sjölander-Lindqvist, A., Risvoll, C., Kaarhus, R., Lundberg Aase, K., & Sandström, C. (2020). Knowledge claims and struggles in decentralized large carnivore governance: Insights from Norway and Sweden. *Frontiers in Ecology and Evolution, 8*(120). https://doi.org/10.3389/fevo.2020.00120

Stoffle, R., & Minnis, J. (2008). Resilience at risk: Epistemological and social construction barriers to risk communication. *Journal of Risk Research, 11*(1–2), 55–68. https://doi.org/10.1080/13669870701521479

Stoffle, R. W., Stoffle, B. W., & Sjolander-Lindqvist, A. (2013). Contested time horizons. In A. Bond, A. Morrison-Saunders, & R. Howitt (Eds.), *Sustainability assessment: Pluralism, practice and progress* (pp. 51–67). Routledge.

Thunold, A., Nøkling, A., & Reksnes, A. H. (2021, January 21). Natur i maskineriet. *NRK.no*. https://www.nrk.no/klima/xl/utbyggere-kjoper-konsekvensutredninger-_-naturen-kan-tape-1.15212845

UN Environment. (2019). *Global Environment Outlook – GEO-6: Healthy planet, healthy people*. https://doi.org/10.1017/9781108627146

INDEX[1]

[1] Note: Page numbers followed by 'n' refer to notes.

A. Sjölander-Lindqvist et al. (eds.), *Anthropological Perspectives on Environmental Communication*, Palgrave Studies in Anthropology of Sustainability, https://doi.org/10.1007/978-3-030-78040-1

D

The manufacturer's authorised representative in the EU is Springer Nature Customer Service Centre GmbH, Europaplatz 3, 69115 Heidelberg, Germany. If you have any concerns regarding our products, please contact ProductSafety@springernature.com

Printed and bound by CPI Group (UK) Ltd, Croydon, CR0 4YY
30/09/2026
02225321-0008